Airports and the
Environment

By the same author
AIR TRANSPORT ECONOMICS IN THE SUPERSONIC
ERA (1967; 2nd ed. 1973)

Airports and the Environment

A Study of Air Transport Development and its
Impact upon the Social and Economic Well-being
of the Community

ALAN H. STRATFORD, B.Sc. (Eng.) C.Eng., F.R.Ae.S.
Air Transport Consultant

© Alan H. Stratford 1974

First published 1974 by
THE MACMILLAN PRESS LTD
London and Basingstoke
Associated companies in New York
Dublin Melbourne Johannesburg and Madras

SBN 333 15177 1

Printed in Great Britain by
WESTERN PRINTING SERVICES LTD
Bristol

Yet all experience is an arch wherethro'
Gleams that untravell'd world, whose margin fades
For ever and for ever when I move.
How dull it is to pause, to make an end,
To rust unburnish'd, not to shine in use! . . .

Tennyson, 'Ulysses'

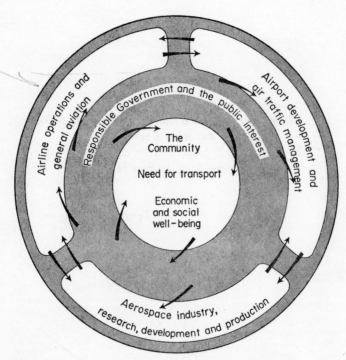

The Community

Need for transport

Economic
and social
well-being

Responsible Government and the public interest

Airport development and
air traffic management

Aerospace industry,
research, development and production

Airline operations and
general aviation

Civil aviation: The wheel of action

Contents

List of Illustrations

List of Plates

Preface

This study has been undertaken on account of the widespread interest being shown today in the development of airports both as national projects and as foci of local activity and concern.

For the most part the views of people can be predicted in the light of their occupations and their place of residence, and many of the opinions expressed so forcibly are quick judgements based on irrationally selected facts.

My aim has been to set out as much of the relevant material as can be comfortably assimilated in a week-end of reading, so that the present status of airports within the community can be broadly established, and the likely changes in the environment over the next ten years can be more fully appreciated.

Aviation is a dynamic business. Changes in the short term will be good and bad. In the longer term strongly favourable changes for the community can be assured.

I have avoided where possible any argument which requires great knowledge of aeronautics, engineering or economics, but the facts and their interpretation are based on my own studies and those of my colleagues which have concerned over forty airfields in this country and overseas.

Mr Colin Waters, M.Sc., B.Sc., with whom I have had the pleasure of working since 1968 on many projects, has contributed Chapter 5, therein providing a simplified appreciation of the theory of aircraft noise and its quantification.

Further reading in the many fields which nourish the subjects of airports and the environment will be facilitated by referring to the References at the end of the book.

A.H.S.

Abbreviations

ADAP	Airport Development Aid Program (USA)
AIAA	American Institute of Aeronautics and Astro-nautics
ALPA	Airline Pilots Association (USA)
ASA	Alan Stratford Associates (Report References)
ATA	Air Transport Association of America
ATC	Air Traffic Control
BAA	British Airports Authority
BAB	British Airways Board
BALPA	British Airline Pilots Association
C of A	Certificate of airworthiness
CAA	Civil Aviation Authority (UK)
CAB	Civil Aeronautics Board (USA)
CNR	Composite noise rating (US Index)
CTOL	Conventional take-off and landing
DOT	Department of Transportation (USA)
DTI	Department of Trade (and Industry) (UK)
EPA	Environmental Pollution Agency (USA)
EPNdB	Effective perceived noise level, decibels
FAA	Federal Aviation Agency (USA)
FAAP	Federal Aid Airport Program (USA)
FAR	Federal Aviation Regulations (USA)
HMSO	Her Majesty's Stationery Office (UK Government publishing house)
Hz	Herz. Frequency Unit, 1 cycle per second
IATA	International Air Transport Association
ICAO	International Civil Aviation Organisation
ICE	Institution of Civil Engineers (UK)
IFALPA	International Federation of Airline Pilots' Associations

Abbreviations

IFR	Instrument flight rules
ILS	Instrument landing system (Airfield Approach Aid)
IMC	Instrument meteorological conditions
ICN	Loading classification number (for runways)
NASA	National Aeronautics & Space Administration (USA)
NASP	National Airport System Plan (USA)
NEF	Noise Exposure Forecast (US Index)
NNI	Noise and Number Index (UK Index)
NPL	National Physical Laboratory (UK)
PNdB	Perceived noise level in decibels
PNYA	Port of New York Authority
RAE	Royal Aircraft Establishment
R Ae S	Royal Aeronautical Society
RTOL	Reduced take-off and landing
R & D	Research and development
RVR	Runway visual range
SBAC	Society of British Aerospace Companies
SBR	Standard busy rate (of airport movements)
SD	Standard deviation
SFC	Specific Fuel Consumption
SNECMA	Société Nationale d'Études et de Construction de Moteurs d'Aviation
SPL	Sound pressure level
SST	Supersonic transport
STOL	Short take-off and landing
TBO	Time between overhauls
TMA	Terminal area controlled air space
TO	Take-off
USAF	United States Air Force
VHF	Very high frequency (radio communication)
VMC	Visual meteorological conditions
VOR	VHF omnidirectional range (navigational aid)
V/STOL	Vertical or short take-off and landing
VTOL	Vertical take-off and landing

The word billion is used throughout in the American sense of 1,000 times 1 million.

Introduction: Air Transport – The Goals and Conflicts

How far have we come?

The technological development of aviation has advanced a long way since the first airfields were chosen in close proximity to our larger towns. When for example in 1928 The Royal Institute of British Architects held a competition for the design of an aerial terminus for London fifteen years hence, the essential requirements included 150,000 square feet for mooring out space, and some 60,000 square feet for garages and workshops. In fact very soon after those fifteen years were out, the existing London Airport site at Heathrow was under development and the 158-acre central zone was being planned. For growth at high rates through a period of rapid technological change has provided the central problem in airport development.

Such development has today become a subject of very considerable importance since we have found already that our principal terminals are choked, and in many areas airfields are totally lacking. We have learnt in fact, rather late in the day, that to a commercial community an airport may soon become as important as other transport facilities, on some of which we have become totally dependent.

What gives this subject its special interest and difficulty, it seems to me, is that no solution can be found purely in technical terms. Other forms of transport are ultimately involved, so are the development of trade and commerce, the location of industry, the influence of commerical aviation on the national economy, the many aspects of regional planning and sociological change – these are all parts of the airport location and development problem.

Hence arrives the complexity, and the necessary contributions from quite diverse disciplines. Perhaps of even greater

importance is our persistent inability to believe in the continuing rapid growth in demand for air services.

I wrote the foregoing paragraphs as an introduction to a
lecture on Airport Development in 1965, and they seem to me
to be entirely relevant today, but on no single line was any
reference made to the subject of noise. In ten years a whole new
generation of air transport vehicles has transformed the scene.

Since the early years of the century transportation has had
an impact on the lives of the great majority of people living in
the developed nations of the world, but only since the 1950s
has it become a significant part of the pattern of that development. Its impact has had both favourable and unfavourable
aspects. As would have been predictable, in some areas,
especially in the vicinity of major international airports, the
reaction to air traffic noise has been very largely unfavourable,
but in some communities, especially in areas where aviation
industries are located, it is understandable perhaps that the
growth of traffic is viewed more favourably, and the negative
aspects of noise and pollution are discounted. Its economic
significance to the overall life of the nation is very rarely
considered.

Such differences of view will inevitably cause conflict, and
this has been evident in all the technologically advanced
countries of the world. The conflict is not restricted to private
and public debate, for riots have ensued in some parts of the
world, notably in Japan, when large-scale airport expansion
plans have been initiated.

The future of air transport

Nor should we give credence to the view that we have seen
the end of the great upsurge of air transport growth and that
world economic forces and the Middle East fuel crisis will
bring this noisy child to rapid decline. All the evidence dis-
counts this.

It is still realistic today to plan for a continuing rise in the
levels of aviation activity in the movement of people and cargo.
It is also realistic to expect a steady growth in the size of the
major airports in their requirements for the handling of
passengers and of the individual aircraft themselves. Inevitably

concentration must continue to be focused upon the developed countries of the world, and on the key routes, where by the mid-1980s it is likely that 1000-seat aircraft will be in full operation. In spite of this there will almost certainly be a healthy advance of civil aviation in the developing countries, especially where tourism is exploited or where major industrial projects can be brought into operation on a world scale.

The author believes that supersonic aircraft must be anticipated by the middle or late 1980s on an ever increasing scale, and that no monopoly will remain for these in Europe and the USSR. Already in the USA reinvigorated research is in train on a new generation of supersonic transports and their supporting technology.

Major airports may well be required to handle 50 million passengers in a year (and probably 20,000 in an hour), but great efforts will also be made to decentralise such intensity of activity by the development of satellite centres for domestic transport services and for general aviation.

There are very good reasons now to believe that the problems of noise, and environmental pollution in the wider sense, should by the late 1980s have become largely controlled, so that only in restricted and carefully planned areas would a residual noise/pollution problem remain with us.

This expectation may appear optimistic to many readers, but a closer insight may convincingly confirm the judgement of the author that technological and operational solutions to the air-transport noise problem should be available in this time scale.

Aims and objectives

A principal object of this book is to identify the key factors in the development of air transport which have caused conflict with the environment, and to discuss the corrective forces which have been put to work. This is essentially therefore a consideration of noise and atmospheric-pollution and of the congestive stress created by aviation where these impinge most seriously upon the population. Thus the metropolitan airports of the world and their environs must be our main concern, and the development of transport aircraft and engines and the techniques of their operation must be a principal theme of our discussion on future prospects for alleviation.

It has become only too apparent in the last few years that the noise from individual aircraft has increased considerably. The number, size and installed thrust of transport aircraft have steadily grown, and the length of runway required has been extended so that the take-off and approach paths of aircraft have been brought into closer proximity to the residential areas which innocently expanded outside our cities in the post-World War II decades. We shall discuss these questions and seek to answer some of the planning problems which arise. We shall also attempt to establish the significance of air transport in the developing economy of a modern state, and look for the short-term and longer-term answers to the problems set by the present dilemma.

Too little is known of the current efforts being made to reduce the noise nuisance in the airport environment. The media indeed have done much to exacerbate the irritation of airport neighbours. Control must indeed be exercised, but the facts need to be aired. Research and development must be continued and extended. Vigilance must be maintained by socially minded individuals without emotional fervour getting the upper hand, and without democratic institutions being swayed by vociferous minority groups. Information on all the major issues must be available, and this has not always been the case.

Aviation is today a key factor in our economic development. To deny its value and to prevent its potential expansion when its recently created bad-neighbourliness is, if properly controlled, of short duration, might be to damage without reason a major benefactor of society.

The plan of the book

In the Frontispiece a drawing is shown entitled Civil Aviation: The Wheel of Action. Three rings are represented. The hub is the community itself. The spokes represent the Government, their sponsorship of development and their responsibility for the public interest. The rim is the Civil Aviation Industry with its three principal components, viz (i) Aircraft and Avionics, (ii) Air Operations, and (iii) Airport and Air Traffic Management. It seems clear that these three elements must be recognised as the fundamental factors in the development of civil aviation, or a quite different basis for the discussion must be established. In airport management, and the development of aviation projects more generally, these factors can be seen to come into action at every major point of decision, and the interests and activity of each may usefully be explored. It will be noted that the chapter divisions in this book have been chosen so as to align with the major groupings discussed above. This might encourage the initiated reader to strengthen his pre-conceived attitudes to the various problems that bedevil the noise issue, but hopefully it will offer him an opportunity to see all points of view, even perhaps to deepen his appreciation of those key areas of conflict which are so difficult to resolve.

The views of the community, the responsibilities of the Government, and the attitude of the three service industries of air transport must now be considered.

The community

Clearly the economic and social well-being of the community at large is a justifiable even if narrow objective for all its members, but the wider interests of the community will depend on its social character and homogeneity, and on its location.

Those close to a major airport must suffer more from noise and perhaps from fear of pollution, and those close to or under flight paths before and after take-off must be more strongly affected than those in remote communities. Those people, moreover, who make frequent or only occasional use of air transport for personal or business reasons, or who make use of general aviation, are likely to be more favourably inclined than those who never use aircraft in any way, and who (even perhaps mistakenly) think that they never will. Here lies a principal dilemma, and at a glance an easily recognised primary cause of conflict.

While standards of social well-being may be open to reason-able discussion and in a given community may be established politically and accepted if only temporarily on a democratic basis, the levels of economic growth required and the need for expanded transport services such as roads and airports is rarely agreed and never accepted without serious heart searching. Views on this subject cross all the political frontiers.

To safeguard the interests of the community it was for

generations considered adequate that the elected representatives in Parliament and in local bodies should plead the case and enact. In recent years, however, independent associations of like-minded residents have been formed. With all kinds of names and objectives these groups have been formidable protagonists, employing legal counsel and technical experts in the public inquiries which have been the required mode for exposing the local objections against major projects within the English Planning System as set in motion after the last war. In this way public participation may be seen to exist, and an opportunity is offered to the Ministry concerned to consider objections, and to seek a more palatable alternative solution; in some cases, perhaps, to find an excuse for delay on a major project even though considered essential by government officers and professional opinion alike.

Responsible government and the public interest

In spite of public inquiry and pressure from groups of like-minded residents, central government is expected to take a long-term wider view with many factors, not obvious to the observer, being taken into account. The interests of British industry and overseas trade, international aspects of air transport development, aviation exports and international engagements at many levels as well as many other facets of the question will be within the purview of Ministers when considering the factors significant to an airport development project. In more recent years the contribution of research and development in improving the air transport vehicle and in depressing the level of noise generated in the airport environs will be the subject of special analysis by the government departments concerned. Judgement on the balancing of the key issues will be required and, while inevitably swayed by political interest, most governments have taken a responsible view and have held a fair balance. The United Kingdom Government decision, now once more in doubt, on the location of the Third London Airport at Maplin was an interesting case of a decision swayed by political and environmental issues which were weighted more heavily by the Cabinet than were the technical-economic conclusions put forward by the Roskill Commission, set up as an independent legal tribunal to investigate the matter.

Decisions by the Government lose much of their credibility through the public debate, the arguments of parliamentary oppositions for political ends, and the methods of television and press who must apparently create conflict in order to hold our attention. In fact most of the major airport planning decisions by the principal air operating nations have been firmly guided and supported by the respective central government and its advisers. Such advisers may often be international advisory and regulatory bodies, such as the International Civil Aviation Authority (ICAO). Behind government decisions lie the specialist departments, sometimes a civil aviation authority, a research establishment, national or major air carriers, the principal airport operating authorities or professional bodies in the particular nation. Independent consultants from the particular state or from a foreign nation may be engaged.

Perhaps equally important is the continuous effort that is being made by every advanced nation engaged in the business of air transport, to research all the relevant areas of importance to the future of air transport in the airport environment. It is the long-term plan for seeking improved solutions to all the problems of airport conflict which the Government and only the Government can sponsor.

In the monitoring of this plan, and in sharing out aspects of work in research and development to industry and to institutions of research, the Government of an enlightened state plays a significant part in resolving the airport conflicts of today.

The air transport industry

As noted above, we can consider this to be subdivided into the three constituent parts – the Airlines, the Aircraft Industry and Airport Management. It is convenient to think of these as the three service industries of commercial aviation. While closely integrated through the vehicle itself they are quite clearly strikingly different in their objectives and in their mode of operation. In seeking to establish a scale of operation acceptable to the community their objectives become unified and it is in this sense that they provide in effect the rim of our wheel of action. In fact they create the action, and they inevitably generate the noise.

The aircraft industry, including the engine and avionics

industry, seeks assiduously to provide the improved or modified vehicles which are required by the airlines to meet new economic and competitive criteria, as well as the noise levels set by governments and recommended by international bodies within the operational time scale.

The airlines, seeking to acquire and operate the improving aircraft types, watchful of the criteria of government in the design stage and in the operating milieu, look to the airport authorities to develop facilities to meet the future requirements of airline and regulatory authority.

Thus we complete the circle, for it is within the province of the airport authority that the public meet the aircraft, either as passengers or as residents subject to the nuisance of noise or afflicted by a planning decision to close a local road so that a runway can be extended.

Can the air operating industries beat into the ground by the end of this decade the problems at issue, or must we wait longer, perhaps indefinitely, for solutions to the conflict with the community which have been created by this vital industry? This is the central question which we have attempted to answer in the pages of this book.

1 The Community and the Contribution of Aviation

'The hostile jibe during the Second World War that
this country was no more than an aircraft carrier should
in the last 30 years of the present century be a source not
only of pride but of economic and political strength.'
—From the Roskill Commission Report, 1971 (Ref. 1)*

It is of importance to recognise from the outset those factors
affecting the community at large which are affected by air
transport development as we move through the 1970s. In
subsequent chapters we shall deal more fully with critical
problems such as aircraft development, regional planning,
airport and traffic management and the programmes for noise
and atmospheric pollution alleviation. We may hope to show
that a viable public air transport system can be developed without restraint upon economic growth if wise political judgement
can contain the raw opinions of pressure groups, by encouraging
public participation in the knowledge of the social and technological options available to us.

Air traffic development

The growth of air traffic through the post-war decades has
been self-evident, but it was the introduction of the turbo-jet
engine initially in the long haul 4-engined aircraft such as the
Boeing 707, the McDonnell-Douglas DC8 and the de Havilland
Comet in the early 1960s which first created the impact. These
aircraft were followed by 2- and 3-engined jet aircraft which
were soon operated in very large numbers.

At this point it is sufficient to note that up until May 1973
no less than 1590, 4-engined and 2700 2- and 3-engined turbo-
jet transports were brought into airline service in all parts of

* For References see p. 153.

9

the world. At that date a further total of 570 jet transports of all types were on order. With this remarkable revolution in air transport came other factors which caused an immediate impact upon the airport situation. The installed thrust was considerably higher than had before that time been available from the propeller-driving piston and turbine engines. Moreover, the straight-through jet exhaust of the early mark turbo-jets caused far more noise than the slow-moving propeller outflow, and the longer runways required brought the aircraft closer to the residential areas around the airports. The higher levels of noise were very quickly noted, and the engine manufacturers, well aware of the seriousness of the problem, made serious attempts to control it.

In the UK, Avon, Conway and Spey jet-engine development at Rolls-Royce progressively offered the aircraft designer increasing thrust at decreasing unit weight for a specific fuel consumption, which rapidly approached the levels finally achieved by the piston engine in its full maturity.

In the most recent development stage the aero engine industry in the USA and in Europe has created several families of newly conceived engines (with high flows of bypass air), to meet the special needs of the civil airline industry. These have provided even more remarkable results in low fuel consumption (per unit of thrust) combined with the lower noise level in the airport environment, which has now become a key criterion for airline and airport operators, and for licensing authorities.

Noise impact

It is not difficult to appreciate that there are four basic factors in the consideration of aircraft noise as it affects people on the ground; from these stem the principal lines of action to alleviate the impact of noise from transport aircraft which are now being taken by government, planning authority, airport and airline management as well as by the aircraft and engine manufacturers themselves. These may be summarised as follows:

1. *The Level of Noise at Source*
   This is generally capable of reduction by improved aircraft and engine design, based on a deeper understanding of the

scientific principles involved. Investment of various kinds is required to produce rapid results; modification schemes for engines may be envisaged and legislative action may prove essential.

2. *The Distance of the Noise*

This is most directly affected by the aircraft design as well as the operational take-off and approach paths and by control of the flight procedures. New types of aircraft with steep climb and approach gradients would make a major contribution in the longer term.

3. *Frequency of the Noise Impact*

This is affected by the control of runway direction within limits set by prevailing winds, by the selection of particular runways and by the diurnal/seasonal variations. Annual growth factors and constraint by curfew will be relevant questions.

4. *Protection of Sensitive Areas*

This is most readily controlled by effective land-use planning, by the proper allocation of land in the noisiest areas and by intelligent use of sound insulation in suitably designed buildings.

Formerly the UK Government could exercise direct control of noise abatement measures only at airports operated by the British Airports Authority. However in the Civil Aviation Act 1971, Section 29, government powers were extended to any UK airport when designated by the Secretary of State for Trade and Industry as presenting a serious noise problem which is not being effectively handled and contained by the management and transport aircraft operators themselves. Moreover, it is laid down quite unequivocally that 'full use must be made of noise abatement measures where these yield a worthwhile benefit. But measures of this type add substantially to airline and airport costs and it would be unreasonable to insist that they should be carried to the point where British Civil Aviation was put at relative disadvantage and traffic was diverted overseas.'

As the UK Department of Trade and Industry has stated in
its progress report on Action Against Noise, 1973, there is
clearly no possibility of shutting down all airports and stopping
civil flying altogether, particularly when a great many people
wish to fly on business or pleasure. Civil aviation makes a vital
contribution to our national economy and also provides jobs
for close on 300,000 people. Indeed more than a tenth of our
exports as measured by value now go by air. Clearly therefore
it has been required that action should be taken by the Government or by local authority to alleviate noise when it is in serious
conflict with the environment, but not to the extent of seriously
restricting the reasonable operation of air services which are
demonstrably of great economic significance to the whole
country.

Few people would dispute the economic benefits arising from
civil air transport whether from the operation of air services,
the utilisation to be made of them, or the industrial value
pertaining to the manufacturer of aircraft engines and ancillary
equipment. As was emphasised in a report on the long term
needs of aviation in the USA published by the Aviation Advisory
Commission in early 1973, the absence of civil aviation would
'clip the nation's pocket book, reduce the standard of living,
circumscribe its horizons, narrow its options and severely in-
hibit the nation's defence capabilities'.

The Chairman and members of one of the most active aircraft noise committees fighting against air traffic development
at London Airport (Heathrow) recently stated that they and
their fellow citizens are

> not in any way opposed to air transport as such, and that
> they fully realise its importance in the national economy.
> Nevertheless, the members of the Action Committee were
> mindful of the fact that at all times the number of people on
> the ground is, and always will be, vastly greater than the
> number in the air, and that the economic and social well-
> being of the community as a whole cannot, in the long term,
> be promoted by sacrificing the majority to any sectional
> interests, however powerful they may be.

It could not be put more succinctly than that.

The need to improve the environment

The views of the environmentalists have indeed been put very forcibly during the last ten years and a consistent argument has been developed which has had a considerable impact upon airport and air traffic management and upon the evolution of air transport services. Based essentially on the human right to live in peace and unmolested by the activities of others except perhaps for very short periods, or when the State is under duress, as in time of war, the environmental preservation movement has grown up in a time of increasing pollution by noise and efflux from all forms of transport, which has itself increased far more rapidly than most other industrial activity.

Coincidental has been a remarkable increase in our awareness of the overall human environment. In Japan, in North America and in Western Europe the changing reaction seems to have been the result of the greatly increased capability of man to affect his environment by pollution or destruction, and perhaps also by reminders of the horrific damage inflicted during World War II, especially by the atomic bombs at Hiroshima and Nagasaki and the subsequent long drawn-out struggles in East Asia.

It is unfortunately the case that much of the so-called environmental pressure arises from motives of short-term interest, when it may be advantageous for individuals to argue against a motorway or an airport project because it is feared that it might effect the value of personally owned property. This may indeed be an adequate reason for opposition, but is surely a doubtful reason for taking an active interest in promoting the diversion of a given project to another route or location, which may effect others in a similar way.

Thus the need is for clear appreciation of the motives behind the opposition of community groups to any major airport project, and the fullest possible understanding of (*a*) the true needs of environmental protection, and (*b*) the fair requirements of private property protection.

It is unfortunately a fact that the finer principles of environmental protection cannot be creatively demonstrated in many projects for environmental improvement. What can be demonstrated however is the work of far-sighted local authorities in

maintaining and improving many areas which had suffered
decay in the last fifty years, and had lacked the benign land-
lords of the eighteenth century to preserve and improve. Thus
the interface between the community and technology is so
often one arising from a conflict between traditional patterns of
living and progress in industry and transport. Airport development has seen its fair share of such conflict. While government
action has to a considerable extent moved in so as to improve
the more blatant cases of environmental damage through its
various agencies, it will always appear sluggish in action to
those who have a more deeply felt interest in the overall environment itself or, in some cases, in the value of their own
property.

It has been argued in the UK that the Department of
Trade, which is the official guardian of the public with regard
to aircraft noise, is also concerned to expand the aviation
and air transport industries. Nevertheless it is the Department of the Environment which is responsible for overall
planning control, and once a project becomes a critical issue
it is called in by the latter Department for scrutiny and perhaps
for public inquiry, so that an overall control of the ground
facilities to be provided is in the final outcome maintained
when the magnitude of a case appears to justify it.

More valid would appear to be the objection that in 1920
the right of the UK citizen to sue on account of aircraft noise
was withdrawn. Such protection for the infant aviation industry
was at that time believed to be justified, and it may be argued
that in the subsequent fifty years the air transport industry
should have been able to reach maturity and fend for itself
economically without inflicting injury upon the adjacent community.

One important factor must, however, be borne in mind. The
jet transport aircraft, the principal cause of the modern airport
noise problem, is in fact no more than twelve to fifteen years old.
The problems have hit us hard and fast, and relatively recently.
Legislation will in due course protect the citizen in most
countries, but a breathing space is required for the aircraft and
engine manufacturing industries, and the air transport industry
itself, to solve the problem of airport noise. This problem is
even younger than the jet transport aircraft, and the sources

of government and industrial research must for some years still be even more energetically deployed to redress the inbalance which now weighs upon the community and on the quality of its environment.

The air transport industry

It is estimated that world-wide employment in the civil aerospace and air transport industry was in 1972 in excess of $1\frac{1}{2}$ million people. In the USA alone threequarters of a million were employed (Ref. 2).

In the UK a total of 80,000 were directly employed in air transport alone in 1972, and of these 65,300 were employed by British and foreign carriers operating in the UK. According to estimates of the Air Transport and Travel Industry Training Board only 7100 were employed in British airports in 1972 a total which is likely to grow only modestly to 7900 by 1974–75. It is possible to show, however, that approximately 300,000 persons were directly or indirectly employed in the UK in 1972 in the civil air transport industry and in industries related to the production, operation and exploitation of commercial aviation (Ref. 3, 4).

There is little doubt, however, that further growth in employment in the air transport industry will be quite moderate over the next five to ten years. Thus a major increase in productivity (output per capita) is to be expected. Not all forms of transport have had a good recent record of increasing productivity, but the steadily increasing potential economy of the transport aircraft has notably contributed and should for at least a decade continue to contribute to a steady annual increase in air transport productivity in spite of increasing costs and the narrow margins of existing civil airline operating profits.

Growth of civil aviation

Though many obstacles have interrupted the natural development of air transport during the last few years, and the Middle East fuel crisis and the economic problems created by inflation are only two of the most critical, the rapid development of air transport needs little proof today. It is accepted by everyone whether actively engaged in civil aviation or not.

The tables and charts below summarise the main features of

the expansion. Fig. 1.1 indicates the growth of world scheduled traffic in passengers and in cargo, expressed as tonne-kilometres, over the last twelve years. Further air traffic on charter flights, and in private executive aircraft, is also significant and growing ever faster.

The figures show the overall result of the ICAO member states operating on all international and domestic airlines. Of the total shown for passenger traffic about 60 per cent is now carried by air lines registered in North America. The rates of expansion has been such that the traffic has doubled in about five years. This approximates to a 15 per cent annual growth (Ref. 5).

The freight, mail, and passenger growth are shown since we are inclined to forget that cargo growth has become as great as, and is often greater than, that of passengers, and for that reason it is increasingly influencing overall airline decisions on re-equipment policy. It may be recognised that the rate of passenger traffic growth has been tending to decline in recent years, although only marginally, taking full account of the last five-year period. It should be noted that cargo may be carried in predominantly passenger aircraft, along with airmail, but an important proportion of air cargo is being carried on all-cargo flights in jet aircraft of types similar to those used for passengers. The contribution of traffic in the Soviet Union is shown by the traffic jump in the records for the year 1970.

Now there is a considerable difference in the rate of advance in different territories and in the areas within one territory. Moreover, the various fields of air transport within one country have developed at very different rates. These varying rates of growth arise through the contrasting standards of economic development in the countries concerned, the geographical situation of the country, the availability of adequate systems of transportation and the degree of support that is provided by the Government for capital airport installations and other facilities and services. Many other factors also arise. In no other field of transport is political incursion more potent for good and for harm than in aviation. The strength of aviation and its weakness lies in the international areas where it has operated with the greatest effect, except in the USA and the USSR whose extensive domestic networks have many of the

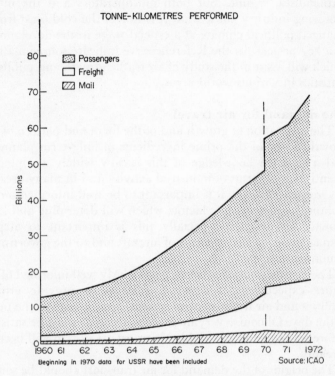

TONNE-KILOMETRES PERFORMED

Legend:
- Passengers
- Freight
- Mail

(y-axis: Billions, marked at 10, 20, 30, 40, 50, 60, 70, 80)
(x-axis: 1960 61 62 63 64 65 66 67 68 69 70 71 1972)

Beginning in 1970 data for USSR have been included

Source: ICAO

Fig. 1.1 World Airline Traffic, Scheduled Services, 1960–1972

characteristics of the international services operated in Europe
and Asia.

Civil aviation may gain sound support from a wise and far-
sighted national government or an extravagant boost from a
nationalistic regime, but both air operators and the manufacturing industry know what it is to feel the cold blast from a
changed political climate at a crucial stage in the development
of a key project. In the References we indicate source material
which will assist in the study of air traffic growth and published
statistics in various world areas.

The demand for air travel

The estimation of growth and of the form and pattern of this
growth is one of the prime ingredients of long-term planning,
and a detailed knowledge of this is now widely accepted as
essential in all forms of industrial activity and in many areas of
government. Clearly it is important to be well informed on the
technical and economic factors which will determine and limit
growth. In aviation especially this is important to airport
management, to the operator of aircraft and to the government
administrator.

The last-mentioned must be particularly well informed of the
future expectation of demand if he is to provide ground
facilities and air traffic control systems appropriate to the needs
of the day. Of course very many requirements must be satisfied
before an economically viable air transport system will become
a reality.

The origins of the demand for air transport should be sought
in relation to all forms of transport and in relation to the demand
for other goods which may be desirable to a population with a
certain level of purchasing power. It is therefore important
that the type of passenger and the cargo commodities whose
movement is forecast should be established so that the elasticity
of demand in respect to the fare and freight rate level as well
as to the other variables such as speed, comfort and safety may
be at least roughly estimated.

It is no simple matter to assess these accurately and to
establish the principles upon which growth and development
depend. Much research has been carried out into the motiva-
tions for the air passenger journey, and important recent US

studies may be referred to. It is to be noted that the work conducted in the US is not necessarily applicable in other countries. The results must be tested for the special conditions existing in that country with its great distances, high-class long-distance roads, high GNP (promoting business and tourist travel) and high marginal incomes (promoting the repetitive use of scheduled and private air transport).

The large majority of people make a journey for one of three reasons:

For business, including government and official journeys.
For private, personal or family reasons.
For tourism.

These three reasons are determined by rather different factors. Business reasons are not self-selected, and although senior personnel and highly-paid employees travel more than middle- and lower-grade staff, nevertheless the income of the traveller seldom seems to influence directly the frequency of his journey or the mode or class of his transport. In many areas business represents about 70 per cent of all air journeys.

Private journeys are generally of an infrequent nature and the use of air transport is rather more dependent on the income of the traveller and, of course, on the country in which he travels, as well as the travel facilities provided.

Tourism is now a major source of air revenue. It is very much more sensitive to the air fare than the other air travelling sectors. This is accentuated by the tendency to travel in family groups, and because transport is only one part of the total expenditure to be incurred. It is sensitive, to a greater or less extent, to the other independent variables of air transport. These are speed, comfort and convenience, and safety.

Air transport as a boon in the mid-twentieth century

The jet transport aircraft has become a factor of immense significance in the last fifteen years and it is essential to the study of airports that we recognise the reasons for its success. To the business man there is little doubt at all that the increasing opportunities that have been offered to him for direct scheduled flights to the main commercial and industrial centres of the world has been a key factor in international trade and commercial development. It is clear that the value of speed in transport increases steadily as the value of time increases (Ref. 5), and the essential links between business organisations in all parts of the world become year by year more dependent upon the facility and speed offered by the airlines of the world. Scheduled air services are of essential value to governments and officials organisations of all kinds no less than to business firms. Though we may discount the economic value of the private use of scheduled air services, they do provide an important social link between many communities and the ties between the UK and Canada, Australia and New Zealand cannot alone be maintained by charter flights and by other cut-price service arrangements, valuable though these may be for long-planned journeys and holidays in the English-speaking Commonwealth countries.

The UK-Ireland traffic is now largely provided by scheduled air services, and it is important to note that the North Atlantic air traffic is still dominated by the scheduled carriers.

Tourism may appear to be of less importance to the community than the business and government air traffic which was discussed above. However, this is not so. Tourism not only provides recreational outlet of inestimable value to the industrial populations of advanced societies, but it has become a source of revenue of very great importance to many states, both highly industrialised states like the UK, France and Italy but also developing countries like Greece, Turkey and Egypt, Indonesia, Thailand and Kenya.

While it is true that the air passenger activity per capita varies very widely between states, and between different parts of the world (Table 1.1) it is nevertheless true that considerable economic benefits accrue to nearly all states from the operation of air services, especially those on international sectors.

Other favourable factors mainly political encourage States to provide air transport and these are even more difficult to quantify.

Britain and the other countries of Western Europe for more than three centuries controlled the trade routes of the world, from their central and strategic maritime positions on the north-west coasts of Europe. The development of air services, however, has thrown wide open the air lanes of the world, which have become concentrated along the busiest trade sectors both international and domestic with scant respect for coast or mountain barrier, though paying rather more respect to the ideological boundaries which have been erected since 1918. Though the USA may generate about 60 per cent of the world's air transport movements on internal and external flights, nevertheless Britain has an immense economic advantage through the position of the London Airports at the hub of the complex North Atlantic-European-Middle East-African traffic patterns.

TABLE 1.1. *International Air Passenger Activity per Capita, 1969*

	Passenger-Kilometres
North America	162
Europe	135
Latin America and Caribbean	33
Middle East	46
S.E. Asia and Pacific	14
Africa	16

Source: Ref. 6.

The value of air transport to the UK economy

A significant report on the subject was published by the chief economist of the International Air Transport Association (IATA) in April 1970. This drew attention to the employment incomes and expenditure generated by civil air transport and related industries in the UK economy. Already by 1968 external earnings had totalled £178 million which was an increase of 56 per cent over 1964. By 1972 this had risen to £304 million, making a net contribution to the UK balance of payments of £19 million.

It was estimated at that time, however, that the gross external earnings were in excess of this figure by virtue of over-seas airlines disbursements in the UK, by exports earned by UK civil aircraft and engine manufacture, by expenditure of air visitors to the UK and by aviation insurance business generated in Britain. These total earnings reached £670

million in 1968 (Ref. 7), and probably well over £1000 million
by 1974 (author's estimate).

The following table summarises the figures computed by
IATA for the years 1964 and 1968, and to this the author has
added his own estimate for the year 1974.

TABLE 1.2 *Gross United Kingdom Earnings from Civil Aviation and
Related Activities.*

	£ million		
	1964*	1968*	1974†
AIRLINES			
Passenger Revenue			
Visitors to UK	44	78	174
Other passengers	45	54	80
Freight Revenue	11	24	46
Other	14	22	40
	114	178	340
Overseas Airline payments in UK	29	55	80
Total Airlines	143	233	420

TABLE 1.3. *Other Sources of United Kingdom Revenue from Civil Aviation*

	1964*	1968*	1974†
AIRCRAFT MANUFACTURING			
Civil Export Earnings	49	142	300
AIR VISITORS TO UK	124	195	350

Source: *Ref. 7, and †estimates made by the author.

It is of interest to note that foreign tourists, of whom 65 per cent
arrived by air in 1973, are spending at the rate of £750 million
in the UK, and by the year 1980 this figure is expected to
considerably exceed £1000 million. Such figures published by
the UK Department of Trade and Industry repay careful
study (Ref. DTI. Monthly Bulletin).

Of the gross revenues acruing to the UK economy as indicated
in the previous table approximately 72 per cent arose from the

activities generated at UK airports. Moreover, the civil aviation linked activities of the UK have clearly been high-growth performers. As was shown above, air traffic growth has been spectacular over more than a decade, both in passengers and cargo, and most current forecasts show a strong continuity in the favourable trends. Tourism is a significant factor in the growth pattern, and not only does tourism boost the flow of dollars and other foreign currency into the UK, but through the income multiplier process it can also be shown to cause high additional expenditure flows. About half of tourist expenditure is on food and accommodation, and this amounts to about 10 per cent of the hotel and restaurant trade turnover in the country.

TABLE 1.4. *Overseas Visitors to the United Kingdom in 1971*

	Thousands	
	All Modes	Air
Holiday	3409	1935
Business	1349	1193
Friends and relations	1269	781
Miscellaneous	946	582
Totals	6973	4491
Overall by air	64·5%	
from USA by air	81·5%	
from West Europe by air	52·6%	

Source: British Tourist Authority Statistics Nov. 1972.

In 1972 about 86 per cent of all business visits to and from the UK were made by air, and for intercontinental trips our own surveys* of business journeys from the UK regions have shown that about 98 per cent are made by air. Fig. 1.2 demonstrates this in the changing pattern of surface and air transport over the North Atlantic during the period 1948 to 1970.

No major change in the pattern of growth taking the trends of a whole, and not viewing individual years as significant is likely to occur.

* ASA Surveys of UK airport developments.

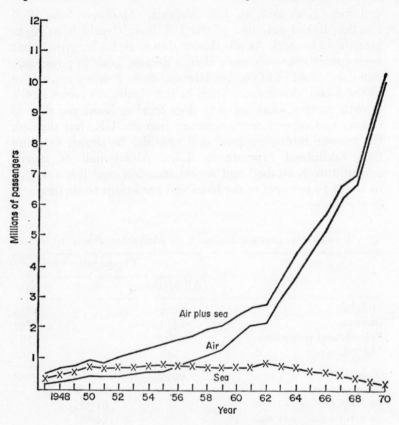

Source: U.S. Department of Justice, Immigration
and I.A.T.A.

Fig. 1.2   Penetration of Air Carriers on the North Atlantic

Net overseas income

The reports of the Air Corporations (combined in 1973 to form British Airways) have shown a consistently favourable balance of international payments. This increased steadily from the early 1960s. In the last published figures of the corporations before amalgamation, for the year 1971–72, it was reported that of BEA total passenger revenue of £131 million, 45 per cent was generated outside the UK, 38 per cent in Europe, and 7 per cent in North America. In the case of BOAC, of the total revenue of £212 million earned in 1971–72,

66 per cent was earned overseas (26 per cent in the dollar areas). Comparable figures were published by the British Airways Board for the year 1972–73.

It is interesting to note that our airports, no less than our airline operators, earn important sums in foreign exchange. The British Airports Authority, for example, at its five airports in the South-East of England and in Central Scotland, earned £11·7 million in the year 1971–72. This was 31·5 per cent of total income and was largely generated at London (Heathrow), where over 74 per cent of the BAA's income is earned.

Overseas airlines operating within the UK record earnings which considerably exceed their expenditure, but on the whole account of civil aviation in the UK, invisible trade creates a surplus which is fairly consistently maintained. Over the period 1963–68, for example, the contribution of civil aviation ranged between 5 per cent and 19 per cent of the total surplus on invisible trade (Ref. 7).

International trade

The value of export goods air freighted in recent years has also increased consistently. In 1968 these goods accounted for 12·9 per cent of total UK exports. This in itself was more than double the value of goods exported in 1965, and at constant prices was probably about double. In 1972, however, no less than 15·8 per cent of UK exports were sent by air. This increasing share is a matter of high significance since it emphasises the fact that the export growth in the commodities most suitable for air freight is at a very much higher rate than the average rate of UK exports. Table 1.5 shows the figures since 1968. Moreover, air freight services are themselves a stimulus to exports and a means of acquiring markets habitually less accessible to British trade.

In 1972 airborne exports grew by 18 per cent over the previous year as compared to 6 per cent total export growth.

In 1972 airborne imports grew by 19 per cent over the previous year as compared to 13 per cent total import growth.

Markets reached by means of air services may, moreover, be more remote in location, and also more restricted in their duration, such as those for fashion goods, for products on exhibition and for perishables. These markets are made

available more readily and often more cheaply by the use of
air cargo services (Ref. 5). The accessibility of markets by air
cargo services has made major airports a focus for industrial
activity. Thus is generated the demand for houses as well as
industrial sites, and thus is augmented the higher incomes and
living standards encountered in the areas peripheral to most
international airports.

TABLE 1.5. *United Kingdom Trade by Air 1968–1972*

	Percentage of Total Trade f.o.b.	
	Exports %	Imports %
1968	12·9	11·8
1969	13·3	13·0
1970	13·6	13·5
1971	14·2	13·6
1972	15·8	14·3

Source: D.T.I.

We have noted already the increased local economic pros-
perity which follows the development of major international
airports. At Heathrow, where the visible trade handled
already exceeded £2500 million in 1973, the pressure on sites
for industry and residential estates has been continually
increasing. One estimate has suggested that £70 million per
year is fed into the surrounding communities from direct airport activity itself (Ref. 33). How significant this is to the
economic well-being of the South-East of Britain is difficult
to assess but any serious interference with such a successful
business enterprise, with its immense potential growth, would
clearly be calamitous to the nation.

Miscellaneous sources of aviation business

A number of ancillary sources of revenue arise from the
position of the UK in the first rank of air industrial nations.

The London aviation insurance market is one example of
this. Over 40 per cent of world aviation insurance is now

handled in London, with an estimated value of £300 million in 1974. Aeronautical training for pilots, air traffic control officers and engineers is also an important source of business activity which is spread widely throughout the British Isles.

Consultancy services are also a significant item and civil engineers, architects, and air transport specialists are employed throughout the world in the airport situations where traffic growth has demanded new project evaluations and system design, as well as the capability for the supervision of major works. Consultancy of all kinds is an important catalyst for trade and services in all parts of the world.

Civil aviation and the US economy

The contribution of the civil aerospace industry and air transport to the well-being of the US economy has been extensively reported. Essential to a knowledge of the subject is the documentation of the Department of Transportation and the National Aeronautics and Space Administration on Civil Aviation Research and Development, published in March 1971 (Ref. 2). In the report on policy, priorities for the 1970s were given as noise abatement, congestion at airports and air traffic control for which were proposed a short-haul system to relieve congestion, and low density short-haul air services to contribute to national transport goals. At the top of the list is the need for noise abatement research and technology which 'will continue until aircraft noise is suppressed into the background. It is a key recommendation of the present study that until this objective is reached, time phased research goals be established calling for reduction of 10 to 15 dB per decade.'

No nation in the world has made greater use of commercial aviation than the USA. The value of jet transport was recognised there even sooner than in Europe, where the gas turbine was first put to use in aircraft. In terms of time saving to the individual and to industry, air transport has never been questioned in the USA. Moreover, the significant world lead taken by the USA aircraft industry since World War II has had a favourable influence on the deteriorating foreign exchange situation and has made an immense contribution to the defence posture of the US Government in all parts of the world.

Various US agencies have studied the impact of civil aviation upon the economy of the country. Such is the extensive documentation of all areas of the US air transport scene that we can at the most quote only three examples of the outputs from these investigations.

Studies of the impact of new maintenance facilities for American and Trans-World Airlines at Los Angeles Airport showed that by 1975 23,000 new jobs would be created, $86 million in constructive work would be commissioned and $630 million of goods and services would be purchased per year. At the Dallas Fort Worth Airport it was subsequently estimated that 46,000 jobs would be created. The Economic Development Administration of the US Federal Government had by 1971 funded forty-three airports and related projects in various economically backward areas in the USA. These projects have increased further the flexibility of the US transport system, and have brought economic benefits to all parts of the country. One interesting example was cited by the authors of the Research and Development Policy document (Ref. 2). 'The Apollo Program involved 20,000 contractors, subcontractors and suppliers, and almost 400,000 non-government workers in all 50 states. Without air transportation, it would have been necessary to concentrate all this activity around a limited number of scientific and academic communities on the East and West Coasts.'

The contribution of the US civil aviation industry to the gross national product is reported in the recent supporting documentation to be increasing at a rate two or three times faster than the GNP of the USA as a whole.

Employment in the USA civil aerospace and air carrier industries is still high in spite of the general air industrial depression of 1970–72, and productivity is very considerably higher than it is in Europe. Estimated employment in the USA civil aerospace industry was 320,000 and in the air carrier sector 445,000 in 1972 (Ref. 4).

The North American governments view their resources in advanced technology and skilled manpower as major national assets and as crucial to the defensive potential of those countries as to their economic strength.

The value of air transport to other national economies

Fig. 1.1 showed the growth of world aviation in passenger-miles and cargo tonne-kilometres carried. But two factors must be emphasised. First, the growth in international traffic will be far greater than that of domestic traffic so that by the mid-1980s probably 60 per cent of all passenger-miles will be generated on international flights (and a lesser percentage, of aircraft movements, probably about 45 per cent, because of the gradual introduction of larger aircraft). Secondly, there is a wide-ranging variability in growth in the various world areas.

Air traffic figures recorded by ICAO have been used in the following tables, which summarise growth aspects of world air traffic and its distribution. The estimates for 1980–90 are due to IATA and have been published in various papers in 1972 and 1973 (see Refs. 6 and 8).

TABLE 1.6. *World Distribution of Air Passengers Provided by Scheduled Airlines*

| | Millions of passengers | | | |
| | ICAO Actuals | | IATA Estimates | |
	1960	1971	1980	1990
North America	64·6	186·4	350·0	560·0
Europe	21·1	74·7	175·0	420·0
Asia and the Pacific	8·1	41·1	119·0	294·0
Other Regions	12·2	26·2	56·0	126·0
World Total	106·0	328·4	700·0	1400·0

Table 1.6 will repay study for it indicates the varying growth rates achieved and forecast in the different world areas. Here lies the dilemma for air transport: the low rates of growth in certain areas with the inherent problem of economic sloth; on the other hand the high rates of development and air traffic demand in key metropolitan areas with insistent problems of noise, atmospheric pollution and air and surface traffic congestion.

TABLE 1.7. *Commercial Air Traffic at Major International Airports 1972*

1. *Air Transport Movements*, exceeding 200,000 in 1972

1·1	Chicago	O'Hare	671,000
1·2	Atlanta	City	420,000
1·3	Los Angeles	International	372,000
1·4	San Francisco	"	297,000
1·5	Dallas	Love Field	291,000
1·6	New York	J. F. Kennedy	291,000
1·7	London	Heathrow	257,000
1·8	New York	La Guardia	254,000
1·9	Boston	Logan	245,000
1·10	Miami	International	240,000
1·11	Philadelphia	"	225,000
1·12	Washington	National	219,000

2. *Total Passengers*, arriving and departing exceeding $11\frac{1}{2}$ million in 1972

			in thousands
2·1	Chicago	O'Hare	33,455
2·2	Los Angeles	International	22,078
2·3	Atlanta	City	21,233
2·4	New York	J. F. Kennedy	20,726
2·5	London	Heathrow	18,679
2·6	San Francisco	International	15,514
2·7	New York	La Guardia	14,235
2·8	Paris	Orly	13,552
2·9	Miami	International	12,266
2·10	Tokyo	"	11,800
2·11	Dallas	Love Field	11,633
2·12	Frankfurt	Rhein-Main	11,611

Sources: BAA, Aéroport de Paris, Port of New York Authority.

Civil aviation and the Third World

The preceding tables have indicated clearly enough that civil aviation world-wide has strong individual regional characteristics.

We have dealt at some length with the UK and USA air transport industries because these have become highly intensive, have created the most serious noise problems and are in fact very fully documented. Aviation is, however, of equal significance in relative terms to most states, and to some of the less developed countries has a heightened importance as the principal channel for foreign visitors, who may provide a large proportional contribution towards a favourable foreign exchange. For the more remote countries of Asia, Africa and South America, air transport also provides the dominant cultural and political links with the outside world.

Third world airlines

An interesting study by the Royal Jordanian Airline, Alia, was published early in 1973 (Ref. 9). This set out the significant economic benefits brought to a small Middle East state by the provision of air services, and may be used briefly as an example of imaginative political/economic action. While 15 per cent of imports were brought into Jordan by air in 1971, and 9·3 per cent of all tourist arrivals came by air (generating 30·7 per cent of total tourist receipts), it was estimated that 20·4 per cent of those employed in the manufacturing sector were employed directly and indirectly in civil aviation.

In Jordan, where the outcome of the Israeli wars is still in evidence, air transport has been able to contribute notably to the rehabilitation of the economy through the years 1969–72, and subsequently in 1973–74.

There is plenty of evidence that in such countries the flexibility of air services can play an important part in the rapid redevelopment of resources, in rehabilitation of dislocated regions, and in renewing the links with outside countries when severed temporarily by war or civil strife. Other examples of this situation could be cited from the countries of the developing world (Ref. 10).

The value of an aircraft manufacturing industry

It was estimated for the US *Civil Aviation Research and Development Policy Study* (Ref. 2), that the world market for commercial aircraft during the twelve years 1974–85 will reach $148 billion. The report indicated that the aerospace industry had become critical to the US economy if for no other reason than that it had by 1971 improved the balance of trade of the US by about $3·8 billion.

The US aerospace trade balance fell to $3·3 billion in 1972 (exports $3.8 and imports $0·5), which helped to redress very dramatically the overall US trade deficit of $6·4 billion in that year.

New civil transports realised $1·14 billion in 1972 exports, and new general aviation aircraft $0·13 billion.

Without this industry an overall national deficit would have been reached in 1967–68.

Thus vigorous steps will continue to be taken in the USA to maintain the leading position it has gained in the civil aviation field and especially in the aircraft, engine and equipment manufacturing field. Continuing assistance will be required from the Federal Government to aid in the development of the new technologies which are the basis of civil aviation leadership. Moreover, new mechanisms for US aviation industry-government interaction have been proposed.

The serious probability is that the fuel oil exporting restrictions in the Middle East will restrain the rate of expansion indicated by the above figures. The impact upon world trade and upon the US balance of payments must therefore be of serious concern. It is the author's view, however, that the restraints upon trade will be short-lived, and that by the mid-1970s it will be the cost of fuel rather than its non-availability which will create the most serious problems for the airlines and airport operators. The extent of this restraining influence upon air traffic would be most difficult to predict.

In spite of the dominant US position in aerospace the European aviation industries are yet very formidable. The UK aerospace exports in 1972 rose to an all-time peak of £417·5 million of which approximately half were civil aircraft engines and engine parts, and in the same year French exports reached £300 million.

The other countries of Western Europe reached about £100 million in air industry exports. While these figures do not reach the scale of US production for export they provide a major contribution to the balance of payments of the exporting countries, a large part of which is the civil aviation element, also developed and exploited by the airline operating industries in their respective countries. The significance of British Airways to British civil aircraft exports, and of Air France to the exports of French Aerospace should never be underestimated. Over the past five years the UK aircraft industry has exported two-thirds of its civil production and has maintained a level of productivity higher than other UK industries of comparable labour intensiveness, some of which have required higher levels of capital input.

However, there is little justification today in considering the aircraft industry of one European country as self-contained and important only as a national asset. Once Europe is effectively united as a commercial community an opportunity will exist to consider anew the possibility of permanent international industrial arrangements whereby one or at the most two large airframe manufacturers, and one aero-engine manufacturer, can stand in the first rank as providers of the world's civil aircraft equipment. Not 8–10 per cent as is the case today, but at least 30 per cent of the first-line equipment of the air carriers could then be exported from Europe, and indeed the percentage could well be very much higher because of the important world-wide commercial and historical links which are and will, it is hoped, be still maintained with the emerging nations throughout the world. If the contribution of the US aerospace industry has been so critical to the US balance of trade in recent years so also will the European aerospace industry have a major contribution to make. The advanced technological base, the management and the labour force are all available in Western Europe. It requires imaginative minds to span the national frontiers so that the international aeronautical organisations can be created, because assuredly the problems of finance, technique and marketing will prove soluble with time and patience. At present employment in the European aerospace industries is about one-third of that in the USA and, yet twice as many people are employed in European industry as a whole as compared with that in the USA. Thus the potential for redeployment within aviation in the European area is immense.

Here lies the need to recognise our responsibility towards developing a more thriving and wide-ranging aeronautical activity, no less for the economic benefit of our own and succeeding generations than for the provision of air services for all of us as private travellers and tourists in this decade and the next. Nor should there be any doubt that properly organised and well distributed airports are essential to the development of an effective air transport system.

The spin-off from advanced technology

In the context of this study we may consider 'spin-off' in two senses. In both cases we are concerned with the subsidiary and subsequent assets acquired by industry as a result of the effort invested in the promotion of new ideas or advanced techniques in front-line projects. But whereas in one case (such as miniaturisation techniques) we may hope to use and/or develop these in new and sometimes quite surprising fields, in the other case (such as carbon fibre technology) we hope to make use of these quite specifically and purposefully in the improvement of aircraft/airport operating economy and hopefully also in the easing of the environmental problem.

The spin-off from Concorde has been exceptional, and remarkable advances have been made by the industry as a result of the associated research programme in the UK and France. Examples are the development of materials for operation under high stress in high temperature conditions, and the evolution of equipment for check-out or test techniques applicable to aeronautical and other advanced transport projects. Perhaps the progress made and experience gained in international collaboration on a major industrial project such as this is even more significant in the long term.

Visits to the Paris Aerospace Exhibitions at Le Bourget in the last six years have demonstrated the bourgeoning of the French avionics industry in the wake of Concorde and Mirage, the successful fighter aircraft. Stimulated by Concorde research and development, and supported by the French Governments of de Gaulle and Pompidou with a national expenditure which grew from less than 5000 million francs in 1962 to over 10,000 million francs in 1970, the aircraft and accessory industries of France were not slow to follow through with wide-ranging products based on advances in technology. An immense long-term gain to the air transport and airport industry stems from all these projects and research programmes.

Examples of the spin-off from the USA Apollo space programme have been widely quoted. The author does not find in these examples many obvious cases showing advantages in the airport situation. It is quite clear, however, that a general quickening of technological intent has been created in all

1 Airports can embellish the environment

Dulles International Airport at Washington showing control tower, passenger terminal and mobile lounge used for taking passengers to and from aircraft

2 Newark Airport in relation to New York City
The need for proximity and the environmental risks when achieving it are demonstrated in this view of Manhattan from the south west

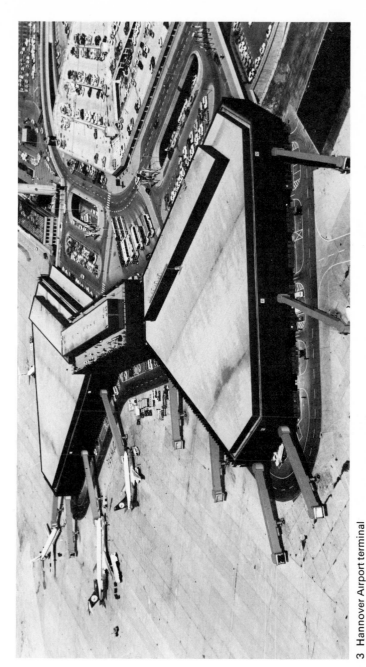

3 Hannover Airport terminal
A view of this recently inaugurated terminal gives an impression of the complex activity on both the air side and the ground side of a modern airport

4(a) Noise abatement at Zurich Airport
A view of a noise-dampening installation at Zurich Airport being used on a
Boeing 747 aircraft of Swissair during engine run-up

4(b) Noise measurement near Manchester Airport
Acoustical tests in a house near the airport to establish the effect of glazing
on sound insulation

advanced industrial countries by these major programmes. Many aeronautical projects such as airborne data-link techniques, high temperature material specifications and manipulating processes have been evolved which are available to the civil aviation industry in a time scale considerably shortened by the costly research programmes thus funded.

Airports in the national economy

The economic significance of the international civil airport as a hub or interface in the transport system of a nation has been inferred from the above discussion. In terms of UK international trade, London Airport (Heathrow) has become the third largest airport in the kingdom. Through the airport in 1972–73 travelled 15,730 million international passengers (82 per cent of the airport total) and 419,814 tons of cargo. A profit of £9·99 million (before tax) was earned in that year, the fifth of a series of years of profitable operation since the airport came under the control of the British Airports Authority (BAA), established in April 1966. While there has been much criticism of the increasing level of charges levied at BAA airports, the facts are that these airports have been operated at an increasing standard of productivity (not only due to the increasing throughput), and a steady improvement in the standards of efficiency, safety and public convenience has been maintained.

In providing this major transport focus for the nation the BAA is aware of creating social problems because of the convergence of aircraft over restricted land areas. It is confident, however, that the steps being taken in noise control through abatement procedures, by aircraft design developments and by local zoning of residential and industrial areas will bring the problems under control within a decade. The major capital city airports have an important role of their own as foci of industry and commerce. Through these hubs of international trade there passes in the final count, the economic power which aviation has helped to create and continues to augment in the interests of every citizen.

Political factors

It is by no means a simple matter to segregate the economic and political issues in a consideration of air service and airport

development. The author has quite simply considered as
political those factors acting in the field of development which
are controlled by the Government without professional
economic guidance or operational/engineering insights available
to provide a basis for decision-making. These factors, therefore,
have a strong emotional content. In this sense it is believed that
there are on balance strong political factors at work which are
favourable to the promotion of air transport services by a
modern state.

While such factors were active in the UK in the pre-war era
when overt government support was given to Imperial Airways
for the development of Empire services exemplified in the
Empire Air Mail Scheme, and were no less obvious in Germany
and Italy, in the last decades direct political support has been
more apparent in the emerging nations of Africa and the Middle
East. It is sufficient, however, to note here that the political
factor is still active, be it obvious or more subtle, and that
powerful opinion in most governments is in favour of the
promotion of domestic and international air services for
passengers and cargo as well as the active development of airport facilities even when the overt economic arguments may
be lacking.

Political forces, leaning often upon bilateral agreements
between states, may not accept, however, unfettered air service
development, especially when the capability for promotion lies
largely in the hands of a foreign based airline. The guiding
concept here may be that a fair share of the inter-state business
should be retained by the national carriers acting as the chosen
instrument of the state concerned. It is not restraint of traffic
which is at issue, but the reasonable claim to economic justice
in the apportionment of sector (or route) capacity, and the
acquisition of foreign exchange and the retention of potential
profits.

The agreements concluded between the USA and the UK in
1946 at Bermuda set the pattern for international bilateral
agreements, though many different forms have now been
agreed. The IATA has come to play an important part in such
agreements. It was not, however, until recent years, when the
non-scheduled charter carriers have come to play such an
important part in international tourist services, that the

fullest implications and limitations of the bilateral agreements have been appreciated.

Short-haul services have become an important part of total airport movements, and it must not be forgotten that in most countries these are domestic operations which are no less significant in a political sense.

In the USA local services still require annual subsidy, and in most European countries forms of government support exist. In fact an increasing awareness of the high cost of political intervention has become apparent in recent years, and within this decade it seems likely that subsidy for commercial air service landing and air navigation facilities will be removed, and that a fully self-supporting system of charges will be introduced in most countries. Nevertheless, so significant are air transport services in the national effort that it is inconceivable that government help would not be forthcoming in an economic emergency for any of the major national carriers.

The airport systems, whether nationally organised or privately managed, are no less the subject of political intervention than air services. In some cases direct government ownership ensures adequate resources and funds, or a high position on the list of priorities for overseas loans. In most cases a strong government interest in airports and ground based air traffic control services is rooted in national defence needs in an emergency.

These manifold questions need the most careful consideration and analysis, and we will discuss each in turn in subsequent chapters.

The price of progress has had to be paid in recent years largely in noise from the air and congestion on the ground. Most of the problems can be solved, or at the least alleviated, by the intelligent use of technical, economic and administrative skills. What is without question, however, is the immense contribution of civil aviation in economic terms to advanced industrial nations such as the USA and Great Britain in the early 1970s, and the virtual certainty that that contribution will increase steadily in the coming decades.

2 The Responsibilities of Government towards the Community and towards the Economic Evolution of Air Transport

'Snorri was the wisest man in Iceland who had not the gift of foresight.'—Icelandic saga quoted by C. P. Snow

'And isn't your life extremely flat
With nothing whatever to grumble at?'—W. S. Gilbert

National and local government are central to the problem of balancing all facets of the community interest against the long-term economic needs of the nation and the region. Central government becomes yearly more involved in the issues, and has now in most states fully accepted responsibility for the control of noise from all sources in industry, transport and aviation.

In many instances the mechanism of control is placed in the hands of a local authority or an airport operator, but the overseers in general are central government officers. Initiative by a local authority has in some instances provided an outstanding contribution to the control of noise nuisance, generally in the form of land use control, but so diverse are the required methods for noise alleviation, including the many technical, operational, economic and safety aspects in the case of air transport, that the incursion of government became necessary more than a decade ago.

An important step was taken by the British Government in 1966 when an international conference was called to discuss in London techniques for the reduction of noise and disturbance caused by civil aircraft. Twenty-six nations and eleven international organisations contributed to the discussions and working parties which focused on areas requiring urgent attention.

Subsequently the 5th Air Navigation Conference of ICAO (1967) recommended action which led to a special meeting on aircraft noise in the vicinity of airports, at Montreal (1969). Standards and Recommended Practice for Noise Reduction were first adopted by the ICAO Council in April (1971) and designated as Annexe 16 to the International Convention. However, the principal states active in air transport have not yet adopted the great majority of ICAO practices recommended in this Annexe. Certain states such as the USA consider regulation of the monitoring of aircraft noise on or in the vicinity of airports to be largely a function of regional or local authorities. Moreover, as yet an approved noise exposure index has not been adopted because of the uncertain accuracy and applicability of the indices proposed (Chapter 5).

The basic concepts concerning aircraft noise have been discussed for a great many years. Since the advent of jet transports in significant numbers, the principal factors determining noise nuisance have become well understood, but the relative importance of these has changed with the widening horizons of air transport in the late sixties and early seventies. Hoped-for solutions have receded in some instances, but new prospects for technological breakthrough have emerged. Meanwhile air transport has expanded continually, and the problems have become more acute, thus obliging governments and industry alike to seek new technical and operational solutions and to open up new areas of research. In this chapter it is hoped to give some account of the recent activity in government aimed at the control of noise in the airport environment. A look will also be taken at various national viewpoints and the quite considerable differences in the attitude of local, regional and central government will be demonstrated.

Areas of government activity
The principal areas of government activity in respect to the civil aircraft noise problem are:

(*a*) Monitoring of the intensity and frequency of noise over residential and other specially sensitive areas close to the airports, and along the airlanes to and from the major airports.

(*b*) Promulgation of noise abatement procedures at airports and minimum noise routings on the airlines, in relation to safe air navigation and minimum conflict with economic air operations and pilot fatigue, on the part of private owners and commercial concerns.

(*c*) Promotion of research into fundamental principles and into the development of projects aimed at the reduction of the noise level, so as to protect communities and to aid overseas aircraft sales. Currently in the UK about £1½m is being spent annually.

(*d*) Exchanging information, co-ordinating of research, and making agreements on the international scene. Monitoring ICAO and bilateral arrangements.

(*e*) Certification of transport aircraft to meet the promulgated noise standards. The objective is full international agreement on noise schemes for all categories of transport aircraft and general aviation.

(*f*) Promotion of consultation between major airports and the public. In addition within the UK the Government provides management of certain airports, some of which are owned by the Ministry of Defence.

This wide range of activity touches not only the public at many points but all sectors of the air transport industry as is implied by the frontispiece, which indicates the central role of government in the total range of civil air activity.

Because it might lead to undesirable repetition in the pursuit of our subject we shall not discuss each of the above areas of government activity in equal detail in this chapter. It will therefore be convenient to consider item (*c*) in Chapter 4 where it more naturally fits into the general theme.

The role of the Government and the independent authorities

The role of the Government is today considerably modified by the existence of major independent organisations usually enjoying semi-official status, and deploying powers vested in them by statute for the provision of services and the exercise of limited authority. In UK civil aviation a number of these organisations have been brought into being. They are the:

1. Civil Aviation Authority (CAA).
2. British Airways Board (BAB).
3. British Airports Authority (BAA).

Their influence is crucial, but it is only the CAA which has been able to implement to any considerable extent the activity of government departments in the control of noise. The ultimate responsibility for the protection of the environment must reside with the Government. This is now vested in the Secretary of the Environment.

United Kingdom airport policy

Central to the UK airport development policy is the responsibility handed to the CAA by the Civil Aviation Act of 1971, and by the document of policy guidance issued by the DTI (Cmnd. 4899 of 1972), to develop an overall national airport policy, and to advise the Government on facility planning and provision. The evolving system owes much to the debates on the report of the Roskill Commission, and to the lessons learnt from airport studies in the UK over the last ten years. No single national airport plan will be evolved, but a piecemeal evaluation is being carried out, largely because it allows more rapid solutions to be found and because it is unlikely that a unified national policy could be pursued without immense pressure. There is, moreover, the serious risk that an overall national plan would prove to be wrong or unworkable before it was properly in operation.

In fact a pragmatic policy is being pursued. Encouragement is being given and restraint brought to bear on provincial airport authorities as well as on the BAA in order to promote and match the development of air services and general aviation.

The evolving CAA policy is based on the commissioning of regional studies, the starting-point for which was the original massive work of the Roskill Commission. The research upon which the policy is based has been planned as follows:

Region	Commission	Completion Date
The South-East	Roskill Commission Work	1968–71
	General Aviation Study CAA[1]	1974
The South	Snow-Stratford Study	1969–70
The West	Bristol Study ASA[2]	1971
	CAA Study	1974
The Midlands	Atkins-Stratford Study	1973
The North	Metra Study	1974
Northern Region	Study by ASA[2]	1974
Scotland	CAA 'In-House' Study[1]	1973
North Ireland	Belfast. Northern Ireland Government and Snow's	1972
	Londonderry Study ASA[2]	1970

[1] In-House Study by CAA.
[2] In-House Study by Alan Stratford and Associates.

A number of independent studies such as those by airport authorities in Liverpool, Manchester, Teeside, Leeds-Bradford, Luton, Hurn and Exeter have been carried out, and it is the wish of the CAA to encourage these, knowing that the requirement for special category permission to extend runways and to develop approach facilities as well as the need to integrate closely with the national air traffic control system will require all licensed airport authorities to seek approval for major works from the CAA. A control is therefore quite easily maintained.

The CAA are keen to avoid any visible support for the proliferation of airports, being only too well aware that the establishment of air services is a matter for the commercial judgement of airlines, and that airport facilities alone do not generate air services or passengers (Ref. 11).

The diverse ownership of airports does not simplify either the study of airports or the assessment of their future suitability, viability and likelihood of active development, which is of prime interest to all airlines.

At the present time the ownership of the principal licensed airfields from which substantial UK air transport operations are conducted may be divided into the following main groups:

1. The BAA – 5 airports (excluding the recent acquisitions of both Glasgow and Aberdeen).

2. The CAA, Ministry of Defence, and regional governments (e.g., Northern Ireland Government, Channel Island States, Isle of Man) – 16 airports.
3. Single and joint local authorities – 26 airports.
4. Private ownership – 7 airports.

Thus the lack of central planning has encouraged a diverse ownership which inevitably had varying motives for the development of facilities. This situation is most unlikely to be modified by government action over the next ten to fifteen years. The almost certain medium-term policy will be to discourage conflict, to preserve where possible the environment and to conserve national investment.

The intensity and frequency of noise

The basis of all studies of the airport noise problem is the detailed analysis of air traffic movements in all categories and its forecast levels in a future period. Airport authorities no less than government departments will actively study these problems if only to be ready for the future demand, if and when it comes, with the required facilities. The Government is largely concerned with the critical air traffic flows at the major terminals, and such concentrations of traffic as may bear upon air traffic control in the terminal air space and on airways. As noted above studies by independent consulting firms are frequently funded by government departments in the area of air traffic forecasting and airport project feasibility.

The principal areas of analysis are passenger, cargo and transport aircraft movements. These are broken down into scheduled and charter movements largely because of the very different characteristics of scheduled airline operation, and the charter/inclusive tour sector of the business with generally different types of aircraft, payload factors and rates of growth. General Aviation in some airport situations may provide a large part of the aircraft movements, but the noise of individual aircraft (except early types of jet executives) have not yet posed a serious problem in the UK. The real difficulties may lie in the future, if the high rate of growth of light aviation continues. The methods of analysis do not need to be considered here, but clearly the need is for all possible precision in the forecast levels of traffic at the key airport hubs, with keen insight into the types of aircraft that will be operating, and into the future trends in the pattern of aircraft operation in short (hourly) and longer (monthly) periods of time.

In the UK where the compilation of aviation statistics has now been taken over by the CAA there is still today a need for greater detail in the information on air transport movements. This is apparent in the hearings on air service licensing, now also a part of the CAA responsibility, at which quantitative data are lacking in many areas.

In the USA a far more favourable situation exists and the statistical data required by the CAB for the support of a case is of a higher order.

It is a requirement of the CAB that the certificated air carriers should report their financial and operating results in a specified form for regular monthly, quarterly and annual publication. In a similar form the ATA (Air Transport Association of America) issue returns which are widely published in the USA and in the international aviation press.

It is certainly very much to be hoped that the action required to be taken in the UK and which has been recommended in a number of inquiries will be implemented by the CAA within the next few years.

Noise abatement and insulation

Experience in the UK has stemmed from the dramatic increase in noise at London Airport (Heathrow), when the first generation of jet aircraft appeared in significant numbers. The problem grew year by year also at other major airports, following some years after the experience in the USA.

In the legislation of most countries, aerodrome owners are protected against actions at court for nuisance in respect to aircraft noise in all phases of operation in the air and on the ground. Thus in the UK aircraft noise was carefully excluded from the provisions of the Noise Abatement Act of 1960 which in fact made excessive noise a statutory nuisance.

However, control over noise abatement measures was taken in various Acts including the Airport Authority Act of 1965, and the Civil Aviation Act of 1971, whereby at BAA airports and at other specified licensed airports where a noise problem is thought to exist, various measures to alleviate noise must be taken. Responsibility for meeting the requirements imposed by the Government rests with the airport authority itself. In the 1965 Act the BAA were required, should the need arise, to make grants towards the cost of insulating dwellings around any of its airfields. Such a scheme was first introduced at Heathrow in 1966 and at Gatwick in 1973.

The Heathrow problem is accentuated by the attraction of many people, not only those connected with aviation, to the residential areas adjacent to the major airports for reasons of trade, housing availability, urban facilities and general convenience. The rapid growth of population west of London in the last ten years has emphasised particularly the problem at Heathrow. Indeed large numbers of families have moved into the airport area, irrespective of the noise levels and disturbance created by the airport, although only a very small number of such householders have availed themselves of the sound-proofing grants available from public sources. Unfortunately it is only rarely that fully sound-proofed housing developments are planned. In 1972 when the Greater London Council proposed a far-sighted scheme for Hounslow Heath, which had been a derelict area for many decades, the full weight of anti-noise protest, including in this case the BAA and IATA, was ranged against it, in spite of the well-planned insulated houses and flats, with interior sound levels contained within reasonable limits.

In this instance a reduction of 30 dB could be readily achieved by the high standard of window design proposed with single glazing. However, certain types of aircraft in the take off phase in easterly wind conditions were liable to fly over the estate generating levels of PNdB which exceed 115 (Chapter 5).

The government decision against this project was based largely on this intolerable level of noise outside the properties which sound-proofing could do nothing to modify.

As the whole community benefits from the airport, it is equitable for the Government, acting on behalf of the nation at large, to provide such compensation as is considered reasonable for those whose homes are affected by the nuisance created by aircraft. At the same time land-use control needs to be more

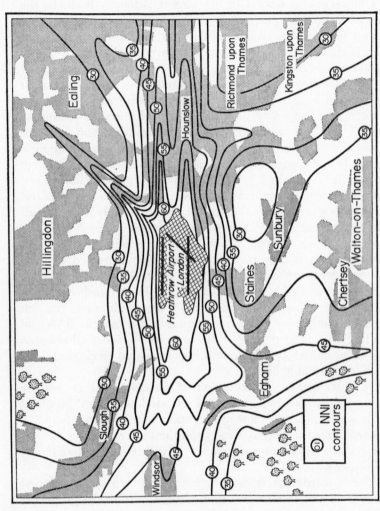

Fig. 2.1 Noise Nuisance Contours around Heathrow Airport, 1970

systematised so that airport management can operate without complaint and harassment, and so that industry and property developers can plan ahead with less restraint.

Within such a system, housing with effective sound insulation needs to be far more fully investigated, and fuller use made of its potentiality in the close-in airport environs. It should not require a closely fought public inquiry and a delay of eighteen months for a decision to be made on options of this kind.

Even in the period 1961–68 a large proportion of the urban and rural district councils around London Airport (Heathrow) were permitting extensive population increases. Reporting in June 1972 to the UK Local Authorities Aircraft Noise Council, the chief engineer of BEA cited areas wholly or partly within the 1970 high-noise contours which had percentage population increases above the national average.

The BEA Survey showed that the population growth rate throughout the area within the 50 NNI contoured zone (Chapter 5) was nearly double the national average of 3.2 per cent for England and Wales as a whole within the same seven-year period. The highest growth areas were:

Langley	30·7% increase in population (1961–68)
East Bedfont	25·6%
Clewer	21·8%
Hayes	17·9%
Stanwell	14·8%
Staines	11·9%
Wraysbury	10·0%

Only those boroughs with an increase of more than 10 per cent are quoted in the above table.

This is a serious indictment of the policies pursued by the local authorities within this area, and should serve as a serious lesson to other authorities still oblivious to the problems and conflicts, which they could do a great deal to avoid.

Housing survey near London Airport (Heathrow)

A survey undertaken by the housing sociologist of the Greater London Council in 1971 studied a sound-proofed low-rise housing estate at Beaver Lane, Hounslow, in the vicinity of London Airport (Heathrow) in order to find out:

(a) Whether the amount of sound-proofing carried out on the estate was adequate for the location.

(b) Whether good interior design and estate layout compensated for the noise levels in the environment.

(c) Whether aircraft noise interfered with the occupiers' enjoyment of their homes and surroundings.

The estate at the time lay within the 55 NNI contour surrounding the airport.

Although only one mile from the east end of the Heathrow Runway 28 Right, the reactions were largely favourable. Single glazed windows of ½ inch plate glass in heavy timber section frames gave a noise level reduction of 28 dbA.

Only 15 per cent of the inhabitants rated noise levels inside their dwellings as noisy. Only 44 per cent noticed it inside their dwellings compared to 41 per cent noticing noise from neighbours.

Noise, however, was very much more objectionable outside the dwellings, where 62 per cent of people rated the environment as noisy or very noisy. Aircraft noise did not appear to interfere a great deal with people's activities. Overall it was found that 47 per cent of the sample were highly satisfied with their living accommodation on the estate, and only 6 per cent were dissatisfied or very dissatisfied.

Such a summary of a complex survey must be incomplete, but it gives strong evidence that a major section of the population in the London area is very likely to be satisfied with well designed sound-proofed houses in a high noise area because of the many overriding advantages which the location can provide.

Airport noise monitoring

An automatic noise monitoring system was introduced at London Airport (Heathrow) in 1966. Noise signals are picked up by the microphones at the monitoring points and relayed to the control point. When the noise level exceeds the critical threshold of 110 PNdB (102 PNdB at night) the signal is taped, and the time recorded. A more modern system will in due course be installed with a more scientific location of the noise-sensitive microphone sites to avoid the practice of evasion which is always possible with a limited number of microphones at well

identified points. Fig. 2.2 shows the layout of the runways at
London Airport (Heathrow) in relation to the take-off flight
paths and noise monitoring points. Most systems at present in
use measure noise only during the take-off phase. It has been
presumed that less variation occurs in the flight path and in the
noise levels occurring in the approach phase, but recent studies
have shown that this is far from the truth, and that landing
phase monitoring is no less important than take-off monitoring.
Moreover, the aircraft noise levels at glide path interception
height may still be very high.*

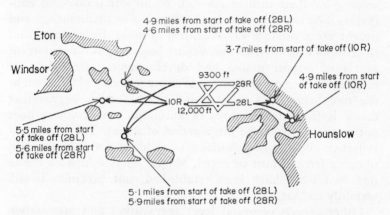

Fig. 2.2   Noise Monitor Locations, Heathrow Airport, London. Noise-
sensitive areas shaded. Distances shown in miles from start of take-off from
Runways 10 R, 28 L and 28 R respectively

An automatic noise telemetering system was installed at
Paris Airport (Orly) in 1973, and the noise monitoring system
at Zurich Airport (Kloten), based on the Danish system of
Bruel and Kjaer, was inaugurated in 1968.

This latter system is followed through by use of a bulletin
which is sent to the airlines monthly. It is understood that pilots
who make a practice of infringement are summoned to Air
Traffic Control for a full briefing upon the detailed noise
regulations at Kloten.

The noise limits have been set both at Heathrow and at J. F.
Kennedy Airport on the basis of the maximum noise level that

* The new Aircraft Noise Certification Requirements do, however, specify
landing and side-line noise also.

is permitted on the nearest boundary of the first built-up area overflown after take-off.

More theoretical methods to establish noise criteria have been attempted. Notable work by the Californian Department of Aeronautics, by Southampton University and by Lough-borough University of Technology has been carried out.

Other sources of noise

Near each end of the runway at a major airport is to be heard a continuous roar from aircraft taxiing and idling while waiting for their turn to take-off. As airport traffic and congestion increases this noise level will rise. The maintenance and freight areas produce other centres of noise disturbance, and their proper location on the airport boundary is an important aspect of airport design and development. The height of engines on the new generation of super jets may place them at the roof height of smaller buildings. A further aggravating factor is the extensive area of apron pavement which does nothing to obstruct the propagation of sound. The significant reflective effect of large surfaces of water and also of adjacent uprising terrain must be noted. Simple relationships for use in our own work have been established, but precision is still seriously lacking.

Other aspects of sound level aggravation and attenuation have been explored – for instance the influence of the topography, and the incidence of wind and temperature inversions – and corrective factors should be applied as requisite. In Ref. 12 a useful account is given of some of these secondary factors which affect noise on the ground with some tentative conclusions on the co-relation between noise complaints and the PNdB and NNI indices at London Airport. This is discussed further in Chapter 5.

 ## The part for regional planning
Regional planning must take full account of the effect of any activity, or use of the land, on all aspects of the environment. This is far easier to spell out than to carry out in practice.

Information of all kinds is lacking, and this applies to forecasts of future use, as well as to the relative impact of different activities on the community and in the longer term upon the quality of life. The planning of a region which includes major industrial, residential and transport hubs introduces problems in a wide range of specialist areas as well as requiring social judgements which can often be no more than personal preferences.

In recent years very serious attempts have been made to reduce the areas subject to personal judgement alone, so that more rational solutions could be found. It must be admitted that this has not been wholly successful, although progress has been steadily made in this field, which has been under study and development for little more than fifteen years.

The airport makes a direct impact upon the community. This is clear enough. But it is vital to recognise that while airports and aircraft have grown and developed so has the community, and many airport-community conflicts have arisen in an acute form because the population in many cases has reached up to the boundary of the airport itself, and may have spread as a ribbon of development along the airport access roads, thereby suffering from air and road noise at the same time. Even in the existing situation residential development is being permitted, without special provision for sound-proofing, in the immediate vicinity of a number of UK airports.

To meet such problems, a number of enlightened communities have established zones for land-use in the airport environs often based on contours of the noise and number index or exposure forecast for future years.

The Surrey County Council, for example, in 1968 introduced a most significant planning document which has been used as a basis for government decisions on acceptable levels of noise. This indicates the framework within which the County Council will consider planning applications for land-use when development of any kind is proposed within the contours of NNI drawn for Gatwick Airport (Refs. 13, 14, and Chapter 5).

Similar techniques have been in use elsewhere, and in the USA widespread acceptance of the concept has been achieved using the NEF or a similar parameter as a criterion for environmental nuisance.

Such methods of control will become more widespread and hopefully will be more acceptable to the public as the relevance of the criteria used can be more surely demonstrated. At the

present time no noise-nuisance parameter has established itself
without doubts. The limitations of the principal criteria are
more fully examined in Chapter 5, but here it may be said that
in the case of the UK NNI system, the social response having
been measured by day in the region surrounding London Airport (Heathrow), it is clearly doubtful whether the relationships
there established can be used in other parts of the UK or
indeed at a later date in that same location.

The NNI contours have also been used by the UK as a
means of providing compensation for residents in the high-
noise areas around the London Airports. A revised scheme for
grants at Heathrow, published in September 1972, offers two
levels of compensation. The defined inner area, within a high
NNI contour, allows for grants up to 100 per cent of the cost of
undertaking sound-insulation work to an approved specification up to a total cost of £360. In the outer area, also broadly
defined by the NNI contours, a grant of 75 per cent of the cost
may be approved. A similar scheme for Gatwick Airport was
introduced earlier in 1973.

We may rapidly be approaching a time when not only will
the overall noise level of aircraft in the various flight phases be
greatly reduced, but engine noise may be so controlled that the
area of noise pollution will be small enough for it to be separated
out as a specifically allocated high noise area of limited extent
and for particular land use, within and adjacent to the airport.

Minimum noise routing

The noise problem should not be thought of as only in the
immediate environs of the airport. Below the air routes allocated
to the airlines a concentration of population may suffer seriously
from the concentration of aircraft on their approach and fly-
away paths. In some regions this has raised acute problems, not
only in seeking alleviation in a particular case, but in finding
an equitable philosophy for selecting aircraft routes. Interested
parties and communities have distinct preferences, and when
power is cut back after take-off so that the subsequent full
power climb away begins at a lower height than would otherwise be the case, the communities below the chosen tracks may
have serious reasons for complaint. The selection of minimum

noise routings (MNR) in effect means the selection of routes
below which the apparent population distribution is the least
dense, taking some account of the height of the terrain and the
needs of the airlines to fly on bearings which do not impose
heavy economic penalties. The question still remains, however,
as to whether a policy of total concentration on the apparent
MNR is fair to the communities thereby exposed, or whether
policies of alternate routings or total dispersion are more
equitable. Total dispersion has severe limitations, and a
considerable degree of concentration is in any event inevitable
in the interests of flight safety because of the need to control
separation, and to use runways in relation to wind direction at
the time.

The UK Noise Advisory Council has considered this fully
and in its report on Flight Routing near Airports (Ref. 15) has
discussed at length the technical problems and limitations
which face the problem of total concentration upon a single
MNR.

Alleviation from a constant aggravation through noise is,
however, always at hand through the vagaries of wind strength
and direction. Nevertheless so important is the problem that
studies are still in progress to clarify the issues. In essence the
problem is to find for an interim period of probably no more
than ten years the most equitable routing policy to and from
the major airports whereby the minimum overall public
nuisance is inflicted. Should we confine the noise to the minimum number of people or should we alleviate the burden by
the dispersal of traffic and the distribution of discomfort?

The key factors in new airport siting and development

A great deal has been written and said on the subject of
airport location during the last ten years, particularly in the
USA and the UK because of the difficulties which have arisen
in the siting of new airports for New York and London. It is
apparent, however, that a large part of it has related to the
vexed question as to whether a new airport is necessary or not.
To those who have worked in this field throughout the period
it has been apparent that much repetition has been indulged in
by those who have, as it were, come upon the subject unawares
and, greatly intrigued, have decided to explore the subject and

to exploit this apparently new field of work. The great appeal of the subject has much to do with its inter-disciplinary aspects, and the opportunity it offers for the exercise of the newer sciences of 'systems analysis' and 'cost-benefit'.

The distinguishing features of the major airport as a subject for site-development analysis are the likely scale of the project, its apparent offer of freedom in the choice of site and the magnitude of the planning options and problems which can be posed. Thus the Third London Airport has been called the largest UK planning decision of the century, and although it is unlikely to incur such heavy capital costs as the UK share of the Channel Tunnel, the overall impact on regional planning in the South-East and the implications for external expenditures could be much greater.

In the first chapter the economic significance of air transport to the nation has been discussed. This is the area for government decisions, either for central or local governments, as the case may be. The need for capital airport projects can be more firmly established in cost-benefit exercises performed upon a short list of selected options. Here is provided the opportunity to assess and balance the favourable and unfavourable aspects of the short-listed sites and attempt to cost those imponderable quantities that are the chief stumbling block of every analyst – the cost of noise, its evaluation for all sorts and conditions of people, the other pollution effects, the loss of agriculture and landscape value, the severance of local roads and the value of the time saved by business and private travellers by location of an airport on one site rather than another.

Today it is wholly accepted that full allowance should be made for the social factors which until recently have been so widely disregarded in the siting of major projects. We have not yet, however, satisfactorily established the quantitative value of so many of the social goods we aspire to retain. The numerical methods of physics and finance have enabled us to design and cost with increasing sophistication the projects themselves, but the contrast in social costing is remarkable.

Unfortunately here lies the opportunity for endless argument and here rests the responsibility of elected government, both central and local, to make the final decisions based on the current assessment of the social values.

The report and papers of the Roskill Commission are essential reading on the subject of the criteria to be used in the siting of a major airport, but we may summarise selected items here for our own convenience, annotating when needful those areas where special requirements for study may apply.

The criteria of airport location

In the existing expansive phase of air transport it is less important to establish the scale of growth of future air traffic than to assess the types of traffic which will form the pattern of expansion and to determine the 'best' of the alternative forms of the infrastructure. How to define 'best' more precisely is our task in establishing criteria. By 'infrastructure' we mean of course the ground and air systems, including airports, ground and air procedures, surface transport modes and interchanges which support the air transport network of airline and general aviation operations themselves.

Traditionally it was the aim to determine airport objectives in terms of engineering and economic principles which established an optimum project among several alternatives, and which excluded the community external to the airport from consideration. Thus, (a) the airfield system, runways, taxiways and approach aids, (b) the terminal complex, buildings for passengers and cargo, administration areas, aprons and car parks, (c) the ancillary facilities, hangars, maintenance areas and approach roads, and (d) air traffic control, might have been the full extent of the engineering investigations with traffic forecasting, revenue/cost appraisals and cash flow studies introduced at the final stage for economic comparisons.

In the mid-1960s, however, a wider view was being taken of the overall situation following earlier studies in the USA and in Europe, although the first of the cost-benefit exercises seems to have been undertaken with respect to the development of dam projects in the USA in the 1930s. We have discussed above how jet aircraft noise grew through the 1960s and how communities spread out from the cities towards the open land which had generally been made available for development in the airport environs. Thus did the viewpoint change.

It is now broadly agreed that the increasing scale and social impact of major national projects such as airports and motorways require that a new definition should be given to the systems, and to the objectives at which they aim. This has caused a considerable change in the work required to be carried out in such exercises when concerned with major projects.

The systems analyst will wish to approach the airport planning process with a clear definition of his objectives. He will wish to know the constraints within which the alternative solutions have to operate, the relationships which bind together the most active factors in the case, and such data from the specialist fields as will quantify his models.

Great value has been obtained from the rigorous methods which have been introduced by the system analysts into the new techniques of cost-benefit, and the many attempts that are still being made to bring nominal social values into the models (or equations) are gradually improving the general level of results, but we are still a long way from achieving an acceptable synthesis of the data from both the technical and the social fields.

The greatest obstacles lie in two areas: (1) the numerical evaluation of the nuisance caused by noise, and (2) the determination of the value of time. In each case a rather wide spectrum of people with varying interests, activities and attitudes is involved, and the evaluations are complex and still very uncertain. Nevertheless because the need today is for the total environmental effects to be included there exists a tendency for the calculated effects to be trusted far more than the facts justify. Indeed for smaller airport projects where jet aircraft movements are unlikely to exceed 10,000 per year within the time scale of the study, the evaluation numerically of the social cost can never be justified and we must resort to quite simple methods of comparative analysis. Housing and school counts within the contours of the aircraft EPNdB, or within the NNI boundaries, at a chosen level of nuisance may be all that is required. This is in total contrast to the determination of an optimum solution to the major capital city airport project for which the combined skills of civil and aeronautical engineers, economists, architects and regional planners, physicists and mathematicians will be required, with all the imagination and skill in project integration that can be provided. Perhaps the leadership of such high-grade specialist teams to achieve the maximum value from their integration is the most essential skill of all.

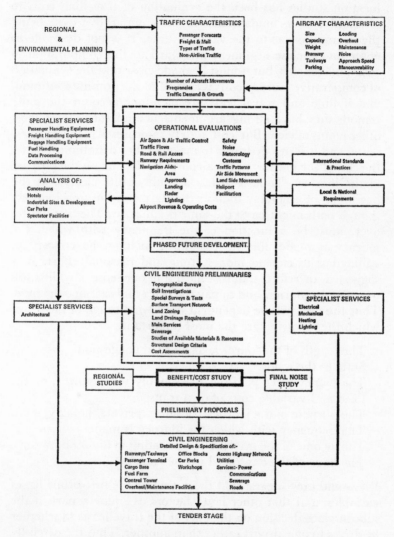

Fig. 2.3 The Comprehensive Approach to Airport Planning
(© Alan Stratford & Associates)

Airport journey costs

A vexed question in all cost-benefit analysis for airport location studies has been the evaluation of travelling costs to alternative sites, fundamental to the comparative costs being the assumption as to the value of time. It is not desirable to enter into a close argument in regard to the computation of costs and benefits, but it must be made clear that the evaluation of comparative journey costs is critical to the optimum solution, and if time and travel costs are of minor account the more remote sites have strong arguments behind them from most other points of view. But 'cost' is to be interpreted in the broadest sense, and it seems desirable to make a fuller definition. Firstly, the cost of the journey to the airport will depend on the purpose of the journey. Time is of less account, quite clearly, at the beginning of a holiday than at the start of a business journey, though both might be to the same destination. The cost, moreover, must be a function of the frequency with which the journey is made, for the cost must include the concept of willingness to expend money, time and personal effort. It is suggested, therefore, that cost means, in our sense, a willingness to make the journey and to pay the price and sacrifice the time. Thus the cost will be dependent on a wide range of factors of which the following are the most important.

The length of the journey, likely road congestion.
State of the roads.
The money price paid out (reasonably well known), or the fixed and variable costs of own transport.
The purpose of the journey (business, private, holiday, etc.).
The frequency with which the journey is made.
The income of the traveller (or his rating in his organisation).
The hour of the day (need to rise early).

We would now suggest that these are just the first prime list of variables, and that other lesser factors can play a part in the subconscious decision on the part of the traveller as to whether he drives to one airport rather than another. Thus the carefully computed exercises in journey cost, which are usually confined (such is the complexity of the overall problem and the 'model') to the journey time, and the values of time to business men and leisure travellers will generally introduce only very broad approximations to a general truth – that we all prefer a short airport journey.

Only in the case, therefore, of a strict comparison of two alternative airport sites, each one to meet a nearly identical traffic flow, will the broad generalisations of the cost-benefit model builder be justified. When the complex and expensive methodology cannot be shown to be justified, more pragmatic methods of analysis and selection may be utilised.

House values

A convenient and measurable yardstick of changes in environmental standard is to be found in house values, either as quoted prices, or prices actually paid, or as values assessed by experienced specialists in the field.

A significant factor in the assessment of the cost of an airport development must always be the damage to the house-owner due to the reduction in value of his house which may ensue from a new airport development, or from an extension of an existing facility. In economic terms the concepts as at present accepted have been spelt out by the Roskill Commission's research team, to whose papers the reader is referred for a fuller discussion.

Some modification to the approach may be perfectly acceptable in a given investigation, the following being the basis on which our own work has been evolved in the most recent UK airport development studies.

The airport management will at all times be well aware of the community noise nuisance contours, on whatever basis they have been assessed, and will be maintaining a count and hopefully some control over the houses, schools and other institutions within the critically high boundaries. Wise managements may be watching house values themselves, but some feeling for the economic mechanism may be of interest.

In effect, house prices or rent charges are probably the most significant measure of the quality of the environment. Most urban development is not dominated by residential buildings, and it is found in practice that urban property values are not a key factor in comparative airport location studies. Moreover, road traffic noise makes it reasonable to exclude the central areas from a consideration of the environmental effects of air traffic acting on its own.

Clearly the market value of a house reflects many factors, some of which vary between people. It comprises the physical aspects of the house and the land on which it stands, the characteristics of the neighbourhood, as well as such questions as ease of access to shops, motorway and railway station and the availability of local schools. All such factors combine in a free market to determine the price to be fetched at auction, or to be realised in rent.

It is the general experience that house property values may tend to rise faster when the further attraction of accessibility to an airport is provided (for employment purposes and/or for use of air services), but that for some particular areas close to the airport or in line with principal flight paths, noise nuisance causes a marked deterioration in the rates of appreciation of value.

The major factors affecting this are noise, fear of accidents, vibration and possible air pollution. Probably the deterioration is determined subconsciously in people's minds by these factors in the above descending order of magnitude. The 'consumer surplus' is that surplus of value defined by economists as the value which might be due to the many advantages adhering to ownership.

In the case of a house, this attached surplus value might comprise environmental factors such as friends, family, schools and other local associations. It is clear that residents in the vicinity of an airport who have enjoyed a high consumer surplus will be the most affected by a serious increase in noise nuisance, and would lose the most from an enforced departure from the area. It must be true, however, that where prospects of employment or of trading are increased by the proximity of the airport some modification to this concept is required.

In this assessment of the cost of deterioration of the environment therefore, it is usual to take account of house prices, often with the assistance of estate agents, as was the case with the Roskill Studies in the Heathrow area.

A detailed sample survey may be required in order to measure the consumer surplus. In the case of the survey by the technical team of the Roskill Commission in rural districts through the

South-East counties an excess on top of the market price of 39 per cent was recorded. It is generally found that the value of high-priced property would depreciate more than low-priced houses, and that the depreciation is higher for all types of property in country areas than in the urban areas where it is more likely that road noise is already a problem. It has been shown, for example, that depreciation is much greater in the vicinity of Gatwick Airport than around Heathrow.

Schools

The cost of noise interference with schools is generally taken at two levels. Firstly, schools very close to new airport sites, or for which the cost of insulation would be prohibitive, could be replaced in a quieter area, and the cost of the new development, less the value of the vacated land, would be assessed.

Secondly, schools in the moderately noisy areas would be sound-proofed so that a maximum internal noise level with windows closed of about 60 PNdB could be maintained. The cost of the sound-proofing measures is then assessed.

A separate assessment is generally required to establish the number of occasions on an average day when the PNdB within the school buildings, with windows open, exceeds 85 PNdB or some other selected noise level. See Chapter 5.

A limit of four movements per hour has been set by some workers in this field as a critical frequency at the individual overflight level of 85 PNdB (Ref. 16).

We have discussed above the critical areas requiring careful investigation in any major airport development, and it may be helpful to summarise these in a summarised form.

Main Areas of Study for Capital City Airport Project

AERONAUTICAL STUDIES
Air Traffic Forecasts for Passengers, Cargo and General Aviation.
Air Traffic Control Studies.
Airport Engineering for Terminal, Runways and Aprons and ancillary facilities.
Meteorology.

REGIONAL PLANNING
Regional Development.

mic Impact of Airport Location.
Development.
Needs of the Airport Worker.
Motorway and Road Systems.
Existing Rail Systems.
Rapid Transit.

ACCESS TO AIRPORT STUDIES
Airport Access Routes and Facilities, Road and Rail.
Value of Time.
Cost of Journeys by all Modes.
Public Transport and the Town Terminal.

ENVIRONMENTAL STUDIES
Noise.
Curfews and Abatement Procedures.
Atmospheric Pollution.
Impact on Residential Areas.
Schools and Institutions.
Soundproofing and Removal Costs.
Compensation Policies.

Noise certification

Recent international planning to introduce noise level certification in new transport aircraft has been generally welcomed as a step forward in implementing current objectives of noise reduction at the source.

The UK Government in the Air Navigation (Noise Certificate) Order of 1970 promulgated noise standards required following international discussions and the recommendations of ICAO. They relate to subsonic aircraft exceeding 5700 kilograms (12,500 lb) of gross weight, fitted with turbine engines with bypass ratios exceeding two. The exemptions, however, should be examined. These include the earlier types of jet aircraft representing a large proportion of existing air fleets. They also exclude all supersonic transport aircraft. Such exclusions are not intended to be permanent. The noise levels are required to be measured in EPNdB:

(*a*) on take-off, at a point on a line parallel to and 650 metres (0.35 nm) from the extended centre-line of the runway where it appears that the noise after take-off is greatest;

(*b*) on take-off, at a point on the extended centre-line of the runway, 6500 metres (3.5 nm) from the start of the take-off run; and

(*c*) on the approach to landing, at a point on the extended centre-line of the runway, 120 metres (394 ft) vertically below the 3° descent path.

As demonstrated by flying trials these noise levels should not exceed those shown in the following table, under the weight conditions noted. For intermediate weights, the noise level should not exceed a linear variation of the logarithm of the total weight.

Maximum total weight authorised of aeroplane	Noise level in EPNdB*		
	At point (A)	At point (B)	At point (C)
272,000 kg or more (598,000 lb)	108	108	108
34,000 kg or less (74,800 lb)	102	93	102

* Defined in Chapter 5.

The noise levels specified may be exceeded at one or two of the measuring points if: (*a*) the sum of the excess does not exceed 4 EPNdB; (*b*) at no measuring point is the excess greater than 3 EPNdB; and (*c*) the excesses are completely offset by reductions at the other measuring points.

In the US the Federal Aviation Regulations, Part 36, have enunciated noise level standards required for the certification of airworthiness, which are closely related to the ICAO recommendations but are higher than the UK requirements, in respect to 2- and 3-engined jet aircraft for which the side-line noise level (case A above) must be satisfied on a line 0·25 nm from the extended centre-line of the runway. Approach, take-off and side-line noise levels are specified in relation to maximum aircraft operating weights, and trade-off is also allowed so that noise levels measured at one of two of the measuring points may exceed the prescribed limits.

Certain allowances were given for 4-jet aircraft applying for certification before 1 December 1969. The regulations exclude SSTs. These noise requirements may prove unduly severe in the case of high-powered aircraft, using steep climb and approach angles with unconventional ILS glide-slopes.

Current jet aircraft operating at noise levels above the requirements for new projects will provide a large part of airline fleet capacity for a number of years. Legislation does not require modification kits to achieve lower noise levels, but some attempts have been made, notably by ICAO and by the FAA, to encourage the retro-fitting of so-called 'hush kits' to achieve improved levels (Chapter 4). Such moves have been viewed anxiously by the CAB, and naturally, also by IATA. Cost estimates have shown excessive financial penalties due to installation costs and aircraft time out of service, and because of the payload/range decreases arising from weight increments.

Due to increasing engine bypass ratios over the period 1965–1974 there has been, and will continue to be, a steady reduction in jet noise from new type aircraft, probably about ½ dB per annum. In 1968 there was a large step favourable to noise reduction because of the introduction of the single-stage fan engine without inlet guide vanes. The further 5–10 dB reduction called for to meet the requirements is possible on most new engine designs but this implies weight and specific fuel consumption penalties, which in 1970 were estimated at over 1000 lb and 3 per cent s.f.c. on the Rolls-Royce type 211 engine (Lockheed 1011) but which in the next generation of transport engines might exact less than one third of this penalty.

There is little doubt that the new requirements will in due course provide considerable improvement in the airport environment, but that airlines and the airline customer will be required to pay increased fare levels for the resulting environmental improvement.

Thus the certification of noise for new types of subsonic aircraft is an important area of government activity in the control of noise which will have far-reaching effects. While apparently not having a direct relevance to the earlier types of jet aircraft and excluding also supersonic transports, the influence of the new standards on the future of these types is already apparent. Airlines' thinking is strongly motivated towards avoiding night flight bans and in achieving the maximum flexibility of operation. Schemes for the modification of early mark engines to achieve lower noise levels are in hand and the manufacturers of the second generation of 2-, 3- and 4-engine jets are striving to seek moderate cost solutions so as to extend the life span of their existing products.

Notes on overseas airport environmental policies

USA

The new Federal Airport and Airway Development Program known as the National Airport System Plan (NASP) is extending the scope of federal aid to airport development on an immense scale. The nation is to spend more on the civil air system in the 1970s than was spent in the previous third of a century. The act not only is providing a healthy financial stimulus for development but is ensuring that adequate procedures are available to safeguard the environment and give opportunity for public participation.

Even so the procedures are complex and are made more so by the USA Environmental Policy Act of 1969 which broadly requires that federal agencies should consider environmental effects in all planning and decision making.

The USA Senate Appropriation Committee has specifically recommended that the National Aeronautics and Space Administration should continue to undertake studies in the fields of noise, pollution and the sonic boom, while at the same time it cut the Nixon Administration request in the fiscal year 1974 for $28 million to less than $12 million.

Congress at the present time reflects the view of the airline industry in the USA that airport noise and reduced take-off and landing problems as well as the economic problems of airline and manufacturer are the areas of true significance rather than SST. Here one may recognise the narrower view of the democratic institution overriding the longer-term concepts, and arguably greater vision of the administration with its wide-ranging specialist knowledge and support structure.

Canada

In Canada the environmental effects of air transport on populations in the vicinity of major airports such as at Montreal and Toronto are little different from those experienced anywhere else. However, the special problems of Canada relate to the wide areas of undeveloped land where the conflict between man and technology is not acute. A major effort to exploit technology was made by the Science Council of Canada, which recommended in 1970 that a national programme of transport development should be inaugurated, based on a STOL aircraft system with aircraft, navigational aids, air-traffic control, STOL ports and supporting ground services.

The STOL aircraft was presumed to be the basis for the system and this was believed to be an advanced technology field in which the Canadians were ahead of other nations.

This lead has probably never been very significant, and the danger lay in supporting a technology which may have had rather uncertain economic foundation. The use of small STOL aircraft in the northern regions has been characteristic of Canadian aviation since the First World War. The dangers of a nation-wide proliferation of feeder services employing small aircraft would be that they would prove costly and almost certainly unprofitable, and very surely would augment the level of aircraft movements at airports and add to the environmental problems of the nation.

At the new Montreal International Airport site near Ste Scholastique, in addition to the 20,000 acres of operational airport land which has been expropriated, a further 70,000 acres has been placed within graded noise zones.

This extensive land area will be controlled so as to minimise population disturbance and land speculation, while directing land use firmly towards desirable occupations, mainly agricultural, which are compatible with airport development and changing requirements through the remainder of this century.

Canada imposed the first complete night curfews on jet operations in North America. These were at Montreal and Toronto.

France

The French Central Administration has established a scheme for the zoning of land peripheral to the major airports in the country based on the forecast situation up to 1990.

Aircraft and engines taken into consideration are those likely to be used in future years, but ground and air procedures are those in current use.

The zones are three in number and are separated, as is the UK practice, by continuous contours. The constraints in each zone are as follows:

	French Index	UK-NNI*	US-NEF*
Zone A	96	56	38

No new contructions, except airport operations and industrial sites. All should be insulated.

	French Index	UK-NNI*	US-NEF*
Zone B	89	49	31

Essential buildings in existing built-up areas only. Agricultural and industrial buildings in fringe areas with insulation.

	French Index	UK-NNI*	US-NEF*
Zone C	84	44	26

Moderate extension of existing built-up areas. Building complexes, schools, hospitals to be avoided.

* See Chapter 5.

Outside Zone C there is no restriction to development on account of noise.

Since 1970 a general decentralisation of the state decisions and control of projects has been noticeable in France. Thus in the management and control of airports, with the exception of those operated by the Paris Airport Authority and other airports of national importance, more authority will reside with the regional state administrations (Prefectures de Region), twenty-one in number. The overall control of noise nuisance, however, is likely to remain with the control administration with guidance from the Technical Service in Paris (Ref. 17).

The French Government is planning to impose an airport tax of one franc on all passengers departing from the three Paris airports of Roissy, Orly and Le Bourget for domestic French flights. The revenue will be used to sound-proof schools and hospitals near Roissy and Orly. This approach may be widely adopted in other countries.

Germany

The German Federal Government in March 1971 passed an Aircraft Noise Act. Noise protection areas encircling the airport boundary, were established and defined as Zone 1, where the German Q Code Index (see Chapter 5) is greater than 75, and Zone 2, between 67 and 75. The Q code is used in Germany as a scale or index for the measurement of noise nuisance. It compares with the US Noise Exposure Forecast so that $Z=67$ is equivalent to NEF$=$30, and $Z=75$ is equivalent to NEF$=$40. Already the estimates for Dusseldorf in 1975 show that the $Q=75$ perimeter will encircle two square miles with the $Q=67$ area reaching 9·25 square miles. A paper by Dr Geert Zimmerman of the Max Planck Institute at Göttingen (Ref. 18) has suggested that as a rule of thumb the size of the areas enclosed by the curves of constant noise nuisance is proportional to the traffic volume measured as take-off weight per hour. The factor for the $Q=75$ area is approximately 0·17 square miles per 100 long tons of weight per hour.

Hewlett-Packard computerised monitoring systems are in use at present at a number of German airports. Such continuous monitoring allows a considerable control of the noise pattern to be exercised by enforcing the established noise abatement procedures. Improved relations with the surrounding communities have been a notable result of the four years of experience at Stuttgart.

Australia

A Noise Abatement Board has been formed within the Operational Division of the Australian Department of Civil Aviation. This department will be responsible for the evaluation of noise problems for the development of standards and procedures for noise abatement. Noise exposure calculations for the future have been prepared for Melbourne, Sydney, Brisbane, Adelaide, Perth and Darwin. Aircraft routings and ILS glide slopes have been modified to minimise noise, and an attempt has been made to influence the design of buildings which are to be erected in the immediate environs of the major airports. A Hewlett-Packard noise monitoring system with ten microphone terminals has been installed at Sydney. Aircraft engine running on the ground is also being monitored. This equipment provides automatic recording and print-out.

A major new airport for Sydney is now being planned. Rationalised site studies on the Roskill team principle were originally intended but pragmatic methods seem now more likely to be used, with very full account being taken of the environmental factors.

USSR

Information on the USSR is necessarily sparse. The state of environmental control in the Union however is very primitive. Most people are unable to express their views, even if they hold any, and no public comment is heard or encouraged. This may have led the aircraft designers in the USSR to neglect the noise aspects of their jet aircraft until the last two or three years, when they have been concerned to sell aircraft in Eastern Europe, in the Middle East, India and Africa. Russian officials have commented on the noise levels of the supersonic TU-144, but little numerical data has been available to allow a direct comparison with Western aircraft to be made.

Overflying Soviet cities is banned. It is very rare for a transport aircraft to be seen near the environs of Moscow or other cities frequented by Western visitors to the Soviet Union.

Japan

Tokyo's present international airport at Haneda, only 15 km south of the city, has posed serious noise problems for a number of years. Already in the early 1970s Japan had been set several of the most severe problems in airport traffic congestion and noise. A report by the McDonnell-Douglas Aircraft Corporation recently estimated that by 1980 about 100,000 million seat-kilometres would be flown by Japanese airlines. For domestic needs alone it was assessed that ninety-four wide-body jet aircraft would be required in that year.

A strong lobby in Japan is pressing the Government to ban aircraft movements after 11.00 p.m. at all international airports in Japan. The airlines naturally wish for operation through twenty-four hours. IATA have put the airline view with the problems of the new airports in mind: these already handle very large flows of international traffic. The sector lengths to and from Tokyo would make the night ban at the new airports a serious economic burden for the airlines to carry because of the dislocation to long distance operations which would be caused in the Pacific area and beyond.

The new airport for Tokyo at Narita (40 km north-east) will be open in 1974 in spite of fierce opposition from radical groups, including farmers and students and local demonstrators. At Osaka in southern Japan, the second major city has still not finalised plans for a new airport. Sites off-shore are being investigated, and immense pressure is being brought to bear by the environmental groups and some political parties with the intention of achieving the postponement of new airport projects.

3 Airport Development and Air Traffic Management: Problems in the Control of Noise and Pollution

'If the State of Hesse comes out again and again in support of the airpor tit does not do so for prestige reasons. It is a matter of relative indifference to us whether the airport takes second or third place in Europe. We are not concerned with making a profit from this business, but we are concerned with the fact that the airport is vital to the healthy economic development of our state. For only this healthy economic development enables us to build schools, hospitals, kindergartens and old people's homes. We do not support this business for its own sake, we support it in order to give our people a better foundation for their life.'—Dr Georg Zinn, former Prime Minister of the State of Hesse.

It is now important that we take a closer look at the airports themselves. It is here after all that the noise problem is created, although far from uniform in its impact.

The wide range of operating factors which determine the ground facilities for air transport and the other sectors of civil aviation must be considered. The variation in scale is immense, and stretches from the small local airfield, with little more than a mown grass strip with the simplest of terminal buildings, to the great international airports where tens of thousands of trained people are engaged in a wide range of activities essential to some aspect of air transport and its infrastructure.

Terminology

The indiscriminate use of the terms 'airport' and 'airfield' sometimes causes confusion. The term 'airport' has come into usage in reference to the large air transport hub, and in the UK especially, where full-time customs facilities are available. ICAO

however, still adheres to the word 'aerodrome' which it defines
as an area of land or water (including buildings, installations
and equipment) intended for use in the arrival, departure and
movement of aircraft. We find in some quarters that 'aerodrome' is viewed in the 1970s as an archaic word; it is certainly
declining in use. 'Airfield' is used with reference to a small
aerodrome, generally without a hard runway or the facilities for
handling transport aircraft, passengers and/or cargo.

Problems of growth

Although smaller airfields, especially those catering for
general aviation, have changed very little during the last
fifteen years, the development of major airports through the
1960s was accelerated by the steady increase in traffic and by
the demands of jet transport for greater runway lengths, higher
pavement strength and greatly increased passenger and cargo
handling capacity. The arrival of the 4-engined jets, especially
the Boeing 707 and Douglas DC8 families of aircraft, was the
principal activating factor. A very extensive programme of
airport building and development was inaugurated, which in
fact still continues.

The facilities already provided form the basis of the existing
airport system, but the new generation of wide-body jets is
already stepping up demands in airport specification. This
consists essentially in the provision of space and accommodation
for handling 'theatre sized' crowds, and in weight and dimen-
sional criteria. Parallel with this, and to a large extent independent of aircraft types and of civil engineering work, is an
extensive programme of airfield approach and en route navigational aids which is in action in all parts of the world.

The fundamental design requirements of the capital city or
metropolitan airports will not continue, however, to be dictated
by the largest of the long-haul aircraft, since congestion is
already a major factor in the costs of airport facility provision
and operation, both to airport owners and to airlines, and
naturally enough congestion is caused far more by the high
frequency movements of the short- and medium-haul carriers.
Congestion arises not only in the air, but with aircraft and
service vehicles in the runway-taxiway-apron system, and in
the terminal area (passengers), and in car parks and on the

roads of access (motor vehicles). Such congestion is for the most part local and although painful to those affected is likely to affect the environment very little.

Under intensive recent study has been the level of atmospheric pollution caused on the airport and in the near lying roads and residential/industrial areas due to the chemical constituents of the engine efflux. This has been shown to be of less significance than originally anticipated and is more likely to be influenced by road traffic and apron motor vehicle movement than by the jet engines themselves. This will be further explored below.

No easement of the problem of congestion is to be anticipated through the 1970s. Indeed attention is being drawn in current studies to further incipient problems due to the wide-body jets because of the expected concentration of movements by these aircraft at the large traffic generating hubs. The avoidance of smaller hubs for economic reasons encourages feeder air movements to connect with the big aircraft. The concentration of such movements by smaller aircraft will increasingly cause problems in the areas of congestion and noise. This wider question of social disturbance through noise and pollution with the more frequent air movements of smaller aircraft may therefore pose greater problems than the long-haul aircraft movement pattern.

Future airport requirements

While in the mid-1970s the airport situation is dominated by aircraft and passenger congestion at the world's major airports, runway and terminal restrictions are already extracting economic penalties at the far larger number of medium-sized airports, where aircraft development has overtaken the tardy planning and investment of many owners and operators. To meet this problem, in a number of countries some form of national airport development programme has been inaugurated. This has the immense advantage of eliminating the less desirable elements of competition and redundancy, but may arouse criticism in some quarters since local enterprise and insight into the special needs of a regional situation might be discouraged.

The US National Airport System Plan (NASP), programmed by the FAA, is of particular interest in this context since it is

likely to be a form of organisation which will be more widely
adopted in the future by states with extensive airport systems
(which may have grown up from diverse origins), especially
where an uncontrolled and redundant hierarchy of airports has
been developed without apparent benefits to the community.

The US NASP comprised in 1971 3238 airports, 27 heliports
and 74 seaplane facilities in public use. In all a further 7400
airfields with landing facilities are in use through the USA,
most of which are in private ownership. As in all countries,
the small airfields and landing strips are of crucial importance
to the air network of the nation, and it is to be noted that in
that country only 60 airports have a runway exceeding 10,000 ft
in length (Ref. 19).

Studies of airport development for the short-term future are
often initiated by the airport owners or authorities who are
under economic pressure to provide expanded facilities to meet
growing traffic demands and the coming generation of transport aircraft. In the less developed countries the urge to attract
air services, more especially to and from the USA and Europe,
has stimulated an airport building programme on a massive
scale. Nearly every major city in the world is today planning
a further extensive development of its international airport
(first extended probably in 1960–65 to suit the early 4-engined
jets), or is seeking a new site to meet the expected traffic
growth of the 1970s.

Although runway length requirements have apparently
reached a limit, parallel runway systems are now viewed with
favour for new airport projects, and the required acreage still
grows with 18–20 thousand acres (6–8 thousand hectares)
being specified for major airports to be inaugurated in the mid-
1970s.

The Palmdale site proposed for Los Angeles requires the
acquisition of 17,000 acres (68·8 sq. km) and the Dallas International Airport now under construction has involved planning
control over 18,000 acres (72·8 sq. km). The scarcity and high
value of land has stimulated interest in off-shore airport sites
when a situation is available within an acceptable distance from
the centres of industry and population. Such locations are also
under active consideration because of air traffic control conflict
with existing major airports near the great cities and because of

increasing public pressure to reduce noise. A recent study commissioned by the FAA investigated new methods of construction for off-shore airport projects. Present interest centres on airports built on piles, those built within polder dams and protected by dykes, and floating airport projects. A Lake Michigan off-shore site is favoured by Chicago; the Foulness-Maplin sands was previously chosen as the site for the Third London Airport, though not without serious doubts. Plans for a seadrome project 10 miles (16 km) off Los Angeles were published in late 1967.

The airport programme in the USA offers a foretaste of the air traffic and congestion which will need to be faced in ten to fifteen years in many other countries. Indeed the level of annual transport movements in the London area in 1970 were equivalent to those of the New York area in 1963. Procedures for the reduction of congestion at the New York airports have been called for. Pilots, planning IFR operations to Newark, La Guardia and Kennedy, are required to file flight plans one-and-a-half hours before their proposed departure time, and are then advised on the delay situation in the New York area. Times of departure (ATD) are also assigned which must be made good within fifteen minutes. Such arrangements must be expected soon in Europe.

Evolution of world-wide airport systems

The growth of airport size and investment has been a natural reflection of the increase in air traffic which has been characteristic of the last twenty years of civil aviation. But the variety of air operations has encouraged many different levels of airport and airfield type with a wide range of facilities for passenger and cargo handling and for the operation of aircraft, from advanced jet types to single-engined light aircraft. The simplest grass airfield now used by general aviation differs in few particulars from the city airports of the pre-war era when runways were rare, and when 500 acres sufficed to meet the requirements of the air carriers.

In the pre-war era capital city airports became important social centres, and Tempelhof, the Berlin airport constructed in 1929, provided restaurants and roof-top observation areas, receiving in its first year of operation over three-quarters of a

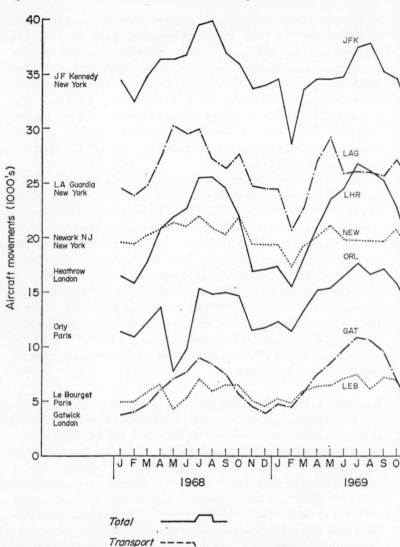

Fig. 3.1 Total Aircraft Movements 1968–71 at the Airports of New York
(J. F. Kennedy; La Guardia; Newark, N.J.); London (Heathrow, Gatwick)
and Paris (Orly, Le Bourget). Showing also for 1970 and 1971 the purely
transport (scheduled, charter, etc.) movements for JFK, Heathrow and
Gatwick (i.e. excluding light, business, military, etc.). *Sources:* Port of New
York Authority, British Airports Authority, Aéroport de Paris

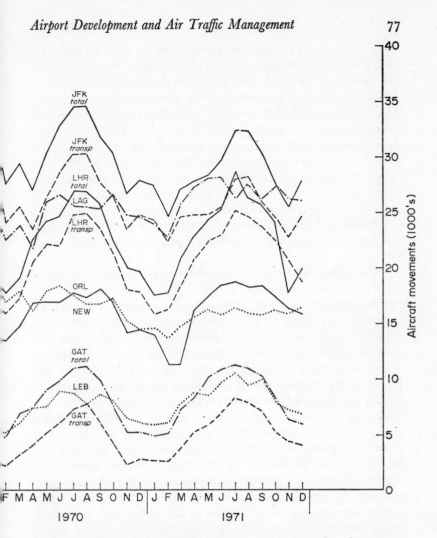

million visitors. Since 1945, however, the story has been one of constant growth, investment and obsolescence.

To set the scale of world-wide airport traffic Table 1.7 indicated the passenger and aircraft movements recorded in the year 1972 for the major airports of the world. The US airports also handle an immense number of general aviation movements.

The annual growth and the monthly distribution of the movements throughout the year is also significant, and Fig. 3.1

shows how the air transport movements as well as the total
aircraft movements have flowed during four recent years in
the premier airports of New York, London and Paris.

Several interesting characteristics are at once to be noted.
Most significant of all is the large scale of air traffic in categories
other than air transport, at the US airfields. This traffic, predominantly general aviation, was growing faster than any other
sector of commercial aviation in the 1960s and is likely to
become a dominant factor in the airport movement pattern
before the end of the 1970s. In spite of some hesitation in the
sales of light aircraft as recorded in 1969 and 1970 this growth
is still strong. Note also the gradual decline in the peak monthly
movements at the US airports as congestion has made itself
felt.

Such a view of the principal world airports, however, gives
no true impression of the world-wide airport situation. For
the countries of the 117 member governments of ICAO an
immense variety of airfields and strips have been provided
which in some areas are only slowly being brought into line
with the requirements of jet transport. In the less developed
areas, as in parts of South America, Central Africa and the
Far East, the preparation of airfields is simple in the extreme,
labour is cheap and the aircraft types in use have few demands
upon the ground facilities beyond a level stretch of ground
with reasonably unobstructed approaches. A beacon, with
primitive lighting of the strip for night landing, is generally all
that is required for the operation of light transport aircraft.

Some elementary buildings for the reception of passengers
and for the airfield personnel to supervise operations with a
tower structure for visual control of air traffic and for ease of
radio communications, may be added when transport movements are on a scheduled basis. From such basic facilities airports move up in the hierarchy to the most sophisticated air
hubs in operation near the centres of our great populations. It
is primarily in these key areas that the immense investments
have been made in ground facilities and the infrastructure of
aids for air transport. In the less developed parts of the world,
which are often in great need of the services provided by
organised air transport, the facilities and the aids may be
inadequately provided. When money is scarce and technical

skills are not available within the country there may be a tendency to provide a fine terminal building and runway, and little of the engineering equipment which the security and reliability of air transport services require far more urgently.

Categorisation of airports

Over twenty years of experience has enabled ICAO to establish standards of operating practice for the design and layout of airports and runway systems. These are set out in Annexe 14 to the Convention on International Civil Aviation. These requirements have been incorporated in the airport licensing and air navigational regulations of most member governments of ICAO (Refs. 20, 22).

The codes of practice relate to runway, taxiway and apron layouts, navigational and airfield approach aids, lighting and obstruction marking. Noise standards are now incorporated in Annexe 16 (Ref. 21). IATA, with the extensive experience of its airline members in operations from airfields in all parts of the world, has also provided extensive guidance material for the airport designer, developer and operator. These latter recommendations have, of course, different objectives and are not mandatory. The most relevant documents are listed in the Bibliography at the end of this book. ICAO has categorised airports in respect to the overall length of the main runway. This parameter, having a reasonably clear relationship to the size, weight and flight safety standards required for the largest aircraft able to make economic use of the airport, and hence to the likely passenger and cargo loads associated, is used to define a code which determines other major design requirements (see Table 3.1).

TABLE 3.1. *ICAO Aerodrome Classification*

Code	Runway Basic Length		
A	2100	m	(7000 ft) and over
B	1500–2100	m	(5000 ft – 7000 ft)
C	900–1500	m	(300 ft – 5000 ft)
D	750– 900	m	(2500 ft – 3000 ft)
E	600– 750	m	(2000 ft – 2500 ft)

The principal factors so determined are:

Taxiway width.
Runway gradients. Average and local slopes.
Width of overall strips.
Transitional slopes for obstruction clearance.
Approach surface slopes. Instrument runway and other runways.
Take-off surface slopes.
Navigational and approach aids.
Obstruction marking and lighting.
Runway and taxiway lighting.

For a full exposition of all requirements in relation to the Airport Category, Annexe 14 of the ICAO Convention (Ref. 20) should be inspected. In the UK the relevant document is *The Licensing of Aerodromes* (CAP 168) published by HMSO. This follows very closely the ICAO code and standards but minor differences have to be considered (Ref. 22).

Runway categorisation

Runways may further be categorised as:

(1) Non-Instrument Runways, intended for the operation of aircraft using visual approach procedures, and
(2) Precision Approach Runways, intended for the operation of aircraft using visual and non-visual (instrument) aids, providing guidance in both pitch and azimuth (yaw) adequate for a straight-in approach.

Instrument Runways are divided further into three major categories defining the minimum meteorological conditions under which safe landings may be attempted:

Category I   Intended for operations down to 60 metres (200 feet) decision height and down to an RVR of the order of 800 metres (2600 feet).

Category II   Intended for operations down to 30 metres (100 feet) decision height and down to an RVR of the order of 400 metres (1200 feet).

Category III   *A*, intended for operations down to an RVR of the order of 200 metres (700 feet).

B, intended for operations down to an RVR of the order of 50 metres (150 feet).

C, intended for operations without reliance on external visual reference.

N.B. RVR denotes Runway visual range.

Master Plans

For the layout of the Airport Master Plan critical minimum dimensions need to be specified and these form the basis of the ground plan arrangements within the defined requirements of the critical aviation and ancillary demand functions.

In our own work we define these functions as:

I. *Terminal Buildings*

Reception	Catering	Other Services
Departures	Other Concessions	Airport Management
Arrivals	Customs	Airport Maintenance
Airline	Concourse	and Servicing
Functions		ATC and Control
		Tower

II. *Aircraft Services*

Aircraft Stands	Fuel Farm
Aircraft Service and Supply	Air Taxi, Club and other
Hangars and Engineering	Flying Activity
Services	Meteorological Services

III. *Airport Services*

Air Cargo Zone	Electrical Supply
Freight Customs	Heating Plant
Industrial Zone	Police and Security
Car Parking	Fire Fighting
Access Roads	

Depending upon the category of airport under consideration the above groupings would be refined or cut so that only the relevant factors apply in the detail required by the project.

In the preparation of the Master Plan a very thorough engineering exercise is called for, involving the combined efforts of many disciplines, including specialist engineers, economists, accountants, regional planners, air transport specialists and systems analysts. Scientists in fields as diverse as noise physics meteorology, soil mechanics, air traffic control and sociology may be involved. The complexity of the problems may be comprehended by a study of Fig. 2.2 (p. 57) which sets out in some detail a work plan for an integrated airport study which was schemed by the author and his associates for an international project.

Airport economics and operations

One essential part of the multi-disciplinary approach to airport planning is the assessment of the air traffic potential in its various forms, such as passengers (business and tourist), freight and mail. It is essential to bring together a team with extensive experience in this field of work, and an effective air transport group must keep in touch with current and future developments, preferably by acting as professional consultants in the air transport field.

At an early stage there must be fed into the studies the requirements arising in the particular airport situation: these may include movements of supersonic transport, the new generation of wide-body jets, the many types of conventional large and small transports and, where called for, executive, private or military aircraft. Provision for adjacent sites for vertical take-off and landing aircraft in the 1980s, and an analysis of the long-term competition from air cushion craft (Hovercraft) or from advanced passenger trains (APT) may become part of such a study. Other forms of transport are now often key factors in the development of air services.

Experience in Hydrofoil, Hovercraft and Container ship economics may need to be brought into play in air transport projects in Western Europe when over-water traffic development is significant.

Two further aspects of airport economics may strongly affect major decisions on airport planning and the detailed specification of the project. These are air cargo development and the potential for industrial activity in the vicinity of the airport itself.

While air traffic and aircraft characteristics, especially in the future decades, are essential factors, studies concerned with road traffic and airport terminal access have in many projects a major contribution to make. The increasing scale of recent airport schemes, and expansion of long-established sites near urban and industrial areas, has emphasised regional and environmental planning. Noise problems have created serious difficulty in most recent major airport studies, and the relevant specialists need to keep in touch with aircraft and engine manufacturers, with particular regard to aircraft-engine noise/ engine thrust characteristics.

Operational evaluations were noted in Fig. 2.2 as central to the phased development of the airport plan. These include many studies specific to the needs of an individual project.

These assessments are brought together with the civil engineering evaluations for the final analysis of benefit and cost. The comparison of alternative projects and investment plans should be based on discounted cash flows.

Airports for aircraft in the new categories

The new generation of wide-body jet aircraft possess certain characteristics which are more demanding than those of their predecessors. These include aircraft length and overall wheel base, wing span and wheel track, turning radius with its effect on runway-taxiway fillet design, and height of the cabin floor with its demand for direct access specialised loading ramps. Increased apron areas and taxiway clearances are required. It is noteworthy that runway length and strength has not been a critical factor in the introduction of these aircraft, whose predecessors set demands never likely to be exceeded.

A higher degree of compactness has been achieved with succeeding generations of jet aircraft. However, we are not in sight of the limit of aircraft size and the many favourable economic factors due to size make it clear that very considerable further increases are likely in the early 1980s.

No major airport development can be assessed without a rigorous analysis of the cost involved and the timing of the expenditure in relation to the revenue earning potential of the project. Thus the cost of the civil works, the land required and the equipment must be ascertained, to the degree of accuracy required by the nature of the study, and the phasing must be determined in relation to the need for increased capacity and facilities.

The growth of air traffic in a given airport situation is

dependent upon a wide range of factors which need to be carefully evaluated in order to determine the most cost-effective plan for promotion.

Feasibility studies

Critical to all studies of airport development is the need to design into the proposals at every stage the essential requirements of minimum noise. Fig. 2.2 showed the many aspects of planning which are involved in a large-scale project.

While environmental questions are only rarely listed explicitly there are few headings which are not likely to include a consideration of questions of noise, pollution or environmental deterioration. Quite apart from the levels of road and air traffic and the noise characteristics of the vehicles concerned we may specify here:

1. The location of residential area and schools in existence (and with planning approval) in relation to the development.
2. The location, direction and design of runways and taxiways.
3. Methods and location of fuel handling and supply.
4. Location of maintenance areas.
5. Location of cargo areas.
6. The layout of aprons.
7. Noise baffles in the run up areas.
8. Design and layout of terminal buildings.
9. Location of peripheral industrial sites.
10. Visual aspects of the Master Plan in relation to the environment.
11 Provision of public recreation facilities in comparison with the original or alternative land-use.

The problem is always to quantify the various schemes that may be under consideration and to avoid judgements based on the traditional codes of engineer and economist to design always within an operational specification so as to provide a solution at minimum cost, subject only to the contemporary dictates of taste and aesthetics.

The concern of airport management

To the airport authorities concerned with the commerical development of major airports, the above issues create the essential day-to-day business which would be neglected only at the risk of serious economic strain or deficiency. No central government or municipal subsidy could for long sustain an airport management which neglected the proper planning and evolution of its facilities and the training of staff. To the environmentalist, however, the development of the airport system is viewed from a quite different direction and he will look to a wider framework of reference – no less perhaps than the total regional environment – for an assessment of the need and significance of the developments proposed. The airport manager for his part will attempt to take a pragmatic view of these questions, knowing very well that he must live amicably with his neighbours, and must seek to resolve as rapidly as possible public debate on the desirability of his plans, but believing quite reasonably that legislation has set out the main desiderata for the required environmental standards, and that common sense can do the rest.

While the airport management must satisfy precisely the requirements of the Government, as specified in the formulated legislation and expanded in the publications issued from time to time (in the UK these are the *Air Pilot*, the Air Navigation Regulations, the Notices to Airmen (Notams) and specific instructions of the DTI and CAA), there are also important areas where considerable freedom for local action exists. In the setting up of the rules for local air traffic, in the use and selection of runways and in the flight procedures, particularly after take-off, the airport manager is, in the general case, the final arbiter.

No airport development plan could be effective without a considerable input of local knowledge and experience from the principal receptacle of this – the airport director himself. Indeed the airport management must be in general agreement with all such plans of development or some very clear reasons must be established for the difference of view.

The situation should be very throughly understood by advisory development engineers and economists, and it requires more than a theoretical appreciation of the points in order to give full value to all the factors in a particular case, which may be unique to the region. The fullest possible appraisal of all the key issues should be an essential part of any such project, and in our own work we have since 1965 introduced regional studies and noise/pollution investigations in every case where serious impact upon an external community can be detected in the pilot study.

When a project has been assessed and the optimum plan has been defined, a wise ownership will consider the early publication of the scheme for public comment if only to avoid the later criticism of the use of arbitrary powers and non-participation. Public inquiry may be called for in many countries, and certainly in the UK, where runway extensions and/or road diversions, with or without land acquisition, is required, the Secretary of the Environment will be likely to appoint an inspector to report to him after hearing the arguments, with full public participation.

The community is generally seriously anxious about the harmful effects of an airport development upon the local environment. The concern is very largely a personal one, but it is idealised to a considerable extent especially when gifted spokesmen in a community energise the feelings of people who might otherwise accept a situation beyond their immediate influence. There is, however, often a very genuine concern over eviction from an expanding site, and fears as to the increase of noise from aircraft (more of it, and from larger aircraft). The influx of augmented road traffic, as well as the urbanisation and industrialisation of existing rural or quiet residential areas also cause disquiet. Without question the fear exists for any qualitative change in the way of life in the area where people now live. Airports are for the most part in areas of sparse population, and hence a dilemma is inherent in the situation. Airport consultative committees can provide a valuable function since by the careful and independent selection of members, useful advice can be offered to management, the committee acting as a pulse-feeler where a restive community has had little opportunity to express its views except by telephone and through the columns of the local press.

The airport consultative committees set up at the airports of the British Airports Authority have been clearly most successful.

With an independent chairman and a membership drawn from local authorities, private industry, trade unions and air-user interests, the free expression of opinion has challenged many of the previously assumed attitudes of management. In 1972–73 members of voluntary anti-noise and environmental groups were invited to join the consultative committees at Heathrow and Gatwick.

The UK Government White Paper 'Protection of the Environment' (Ref. 23) in 1972 reported the setting up of no less than thirty-three consultative committees in the UK. We have already noted, moreover, that people have moved out towards airports, and few airports can be extended today without some conflict with homes and/or institutions of some kind. Perhaps the classical case will always remain, Maplin. Here a UK Government, influenced strongly by environmentalists and against the final advice of the Roskill Commission which was based on a most detailed cost-benefit study, recommended the location of a Third London Airport on remote sands off the South-East coast, only to be attacked more virulently than ever by countless organisations, led by the 'Men of Essex'.

We hear less of the benefits which airports will bring in their train into a region. Higher wages, increases in job opportunity and more visitors to the region and to the country as a whole must surely follow. However, the airport management and its planning advisers must take due account of all such relevant factors, and perhaps on a less complete basis than did the Roskill Commission, give quantitative values where it can, and make broad assessments where numerical answers are not valid.

Night curfews

An area of control which still remains largely in the hands of the airport management is the restriction of night operations. The hours of control can clearly be varied, movements can be constrained and outright banning of all air movements can be imposed.

In November 1971, for example, the UK Minister of Trade and Industry announced a ban on the night take-off of jet transports from Heathrow during the summer months April–October inclusive. Limitations on the number of night movements at Heathrow, Gatwick, Luton and Manchester had already been in operation. The use of take-off control was preferred to the control of landings because of the problem of emergencies, and the need to accept aircraft when directed from other airfields. Subsequently cargo jet take-offs were permitted, with strict limitation on numbers.

Night movements of non-jet transports have not been affected, a factor of great importance to the operators of prop-turbine cargo aircraft through the London area.

Cargo operations are at their most intensive in the night hours, more particularly at the main capital city terminals such as London, Paris, Frankfurt and Chicago; at this time industry brings its products for international distribution, following the working day. Thus special concessions for cargo jets have been sought. At Heathrow, for example, a limited number of night jet cargo take-offs are permitted in the summer months. A number of other major airfields, notably Paris-Orly, Zurich, Frank-furt, Hamburg and Tokyo, have introduced some form of night movement control; in most cases a closure has been ordered.

It has generally been found that the economic penalty to the short-haul scheduled airline of the night take-off ban, or of a restriction on night movements, has been less than anti-cipated. Hardest hit have been the inclusive tour operators who have been dependent on night movements through the peak summer season to maintain the high aircraft utilisation so essential for economy of operation.

The problem for the high intensity tour operator is in-extricably tied up with his equipment policy. While airport managements are likely to encourage very actively the use of high capacity jets in order to reduce the number of aircraft movements and to induce the airlines to acquire newer and quieter aircraft, the imposition of a night ban or restriction would probably prevent the airline from achieving an adequate utilisation in the crucial holiday season. By retention of smaller and older (and probably noisier) aircraft, some airlines could find an answer in a wider spread of operations through smaller airfields where the noise is less disruptive, the environment less sensitive, and for some years at least the night curfew unlikely to be sounded.

An aspect of the night curfew which may be more apparent to the long-haul airline than the short-haul carrier or to the

charter operator is the dislocation which may be caused
inter-continental movements. As much as twenty-five ho
could be lost in the elapsed time between London and Toky
if a seven-hour curfew were uniformly imposed at the key
intermediate points through the Middle East and Far East.

A further problem created by airports which introduce the
curfew would be the congestion in the air traffic terminal areas
and on the ground immediately prior to and after the close-
down hours. It is for this and other reasons that, as noted above,
the take-off ban during summer hours for jet take-offs at
London has in fact already been modified to allow 140 jet
take-offs in that period. Although landings have not been
controlled to date, the position is under constant review by the
Department of Trade and Industry.

Other aspects of noise control

An impressive control over the use of runways in the interests
largely of community protection against noise is maintained at
Heathrow. Subject always to the demands of wind (when of
strength greater than five knots) the direction of use of the
individual runway is chosen so as to affect the minimum number
of people. When possible, westerly take-offs and landings are
made. The parallel runways are brought into play so as to
spread the burden to some extent. For this reason the more
southerly Runway 10 R is used where possible in easterly wind
conditions because of the lesser density of population in the
region of Hounslow Heath.

At Manchester Airport (Ringway), which handles the
fourth largest passenger flow in the UK, the environmental
problems are also considered very seriously. Night jet movements have been restricted, and a scheduling scheme introduced which gives advantage to the new generation of quieter
jet aircraft. Automatic noise monitoring has also been installed.

By the Manchester Corporation General Powers Act of 1971
provision has been made in the interests of noise control for
land purchase and the cost of civil works; powers have been
taken also to enter premises and to allow orders for the sound-
proofing of buildings. Such positive action will almost certainly
lead to a finer control of noise (not only from aircraft), and
assist considerably in the improvement of relations between the

airport committee and the local community. By 1976 it will
be a requirement at Ringway that at least 500 of the 3250
approved night jet movements in the seven summer months
will be with new types of quiet aircraft.

Another twist in the noise problem and methods of defeating
it was demonstrated recently when the Massachusetts Port
Authority initiated a campaign urging citizens affected by noise
from Boston Airport (Logan) to write to congressmen in
support of legislation to effect a jet-engine silencing programme.
This programme for retro-fitting the older marks of jet engines
which comprise still the major proportion of those running
today, is discussed in Chapter 4.

The location of taxiways and ground holding zones well
away from the sensitive peripheral areas is an important part
of airport master planning. Rarely is the background noise of
general airport activity taken into account when assessing the
environmental noise levels. This noise, which comprises airport
motor vehicle movements on aprons and access roads, aircraft
taxiing and idling, aircraft running up in the maintenance
areas, and activity on the main passenger and cargo aprons,
provides at a major hub, a fairly continuous level of noise
which may reach a mean level of 60 PNdB. Sound-proofing
of maintenance areas must be fully investigated and their
location in respect to other activities and in relation to sensitive
peripheral areas must be adequately defined. Moreover, the
need for specialised muffler installations for noiseless full-power
testing must all be fully assessed both to provide for the short-
term proven demand, and also to allow space for further
facilities if and when the less certain future demand should
eventuate.

The conflict between management and the community

It can be recognised from the above discussion that important
but restricted areas of activity are open to airport managements
in their attempt to contain the impact of the noise nuisance
upon the community. It is clearly a requirement that they
develop the airport to the greatest extent possible in the interests
of economic viability, and to this end – no easy one to achieve
in the great majority of cases – the air transport operators must
be encouraged to step up their air services, new air carriers

must be solicited, and facilities of all kinds must be provide in order to sustain the traffic. In his attempt to develop a viable long-term operation, however, the airport manager has now become fully aware of the need to restrain the noise, and he is in no doubt about this if his plans include an extension of run-ways, or terminal buildings which in some countries (as in the UK) will require some form of government sanction and in many cases a public inquiry at which the community can make its voice well and truly heard. Nevertheless the constraints upon management are today as firmly imposed by the Government or Aviation Authority as by the desire to satisfy local opinion, and it is these constraints which offer to the management far less freedom of action than was the case ten years ago. Thus, the management must establish minimum noise procedures and implement reasonable noise regulations acceptable to the Aviation Authority (FAA or CAA) and to the operators, while running a public relations exercise to satisfy the local com-munity. All the while the needs of economic viability will require an increasing level of traffic, and the maintenance of a standard of facilities satisfactory to the air operators and to the travelling public.

Growth of air traffic especially on the more dense inter-national routes has until very recently required more space for runways, taxiways and aprons. Not only have the aircraft and aero engine industries called a halt to this trend but engines of improved noise level have been introduced with the help of research in government establishments. This progress can be expected to continue in the future. We will in the next chapter consider the implication of these changes and discuss their likely effect upon the airport environment during the next decade. Meanwhile we should consider the other aspect of public nuisance which could be generated by intensive airport operations.

Atmospheric pollution

Airport authorities have for some years been alerted to the risk of atmospheric pollution in the airport environment due to the air and surface traffic operating within and in the vicinity of the airport. In fact, owing to the nature of the gas efflux from the gas-turbine engine, which now universally powers transport aircraft, harmful chemical residues are more likely to rise in serious quantities from general aviation (largely petrol driven) and from road vehicles.

Most serious is carbon monoxide concentration which in the vicinity of large capital city airports in the USA and in Europe has been found to have reached levels equivalent to that in dense urban traffic areas.

The contribution of transport aircraft to the total pollution level is made largely in the taxiing operation. Thus the reduction in taxiing time could be an objective in airport design and in the operational methods used.

The US Environmental Pollution Agency (EPA) has proposed standards to limit airport pollution. These will require existing engine design changes during the ten-year period ending 1 January 1983. Emission standards for hydrocarbons and carbon monoxide would require all engines manufactured after 1 January 1976 to achieve a 10–30 per cent lower emission level.

Proposals for modifying operational procedures, such as requiring multi-engined aircraft to taxi on a reduced number of engines, have been made by the EPA but the Federal Aviation Agency has cast doubt on the feasibility of this method of reducing pollution. Trials are in hand for checking ground manoeuvring techniques to give reduced pollution with acceptable operating flexibility.

Clearly the scale of the problem depends upon the relative traffic movements in the main categories. These are essentially:

(*a*) Transport Aircraft (gas turbine powered).
(*b*) General Aviation (still largely petrol driven).
(*c*) Surface Transport (petrol and diesel driven).

Static heating and power plants on the airport may also introduce a problem, though unlikely to be dominant.

The pollution from transport aircraft

The constituent gases from the efflux of transport aircraft jet engines are reasonably well known, and published sources and data from the aero engine industry can provide what is necessary to assess an individual airport situation (Refs 29, 30).

Our own work suggests that the critical operational phases

can be precisely defined, each being allocated a mean elapsed
time for the subsequent computation of gas efflux. Elapsed
times depend upon the aircraft type and the characteristics of
the airport.

The operational (flight and ground) and maintenance phases
are taken as:

A. Engine checks. Start up to taxi out.
B. Taxi out (or push back) to ground hold.
C. Holding pre-take-off.
D. Take-off:
   D1 on the ground,
   D2 airborne to 1500 ft.
E. Landing from 1500 ft.
F. Taxi in.
G. Maintenance, and pre-operational engine testing.

A careful assessment by consultants of the aircraft traffic,
existing and future, and its mixture of types is required. Moreover, annual, peak, daily and hourly movement rates are
required. With airport and aircraft operational times computed,
the pollution emissions can be calculated from the available
data.

The critical chemical products are: carbon monoxide, the
oxides of nitrogen, unburnt hydrocarbons and particulates.
This last item is the visible element or smoke plume which is in
fact objectionable rather because of its unpleasing appearance
than because it is actually harmful to health. We will discuss
below the successful efforts that have been made to eradicate
the visible exhaust plumes from jet engines in the last few years.
Visible smoke has now been considerably reduced, but not yet
entirely eliminated.

A report of the Warren Springs Laboratory of the UK
Department of Trade and Industry was prepared on the air
pollution at London Airport (Heathrow). This concluded that
'the highest values come from road traffic and taxiing operations. . . . There is no evidence that smoke pollution was
affecting nearby areas, but it would appear that during the
winter months urban pollution would exceed that emanating
from the airport' (Ref. 30).

In a study of air pollution from aircraft in Los Angeles the chief air pollution engineer has reported that as jet transport air-craft increasingly dominate the airport environment, there will be a decrease in the emission of other organic gases and aerosols.*

At the time of this study, aircraft accounted for 1·2 per cent of hydrocarbons and other organic gases, 10·4 per cent of aero-sols, 1·8 per cent of nitrogen oxides, and 16 per cent of carbon monoxide in the atmosphere across the county of Los Angeles.

In any comparison of jet transportation with other forms of transport, it may be claimed without fear of dispute that air-craft provide one of the cleanest form of transport. As measured in pounds of pollutants per 1000 seat-miles, gas-turbine-powered aircraft produce less than one-half the weight of pollutant from diesel electric trains, and less than one-fifth of the pollutant generated by new and improved motor vehicles meet-ing stringent efflux requirements for urban areas.

Progress in jet-engine technology has not been restricted to thrust output, fuel consumption, specific weight, noise levels and smoke abatement. The early turbo-jets may still discharge as much as 350 lbs of total pollutant during the taxi and take-off phase of flight. This is of course a considerable improvement on the early piston-engined transports of the 1950–60 era. However, new technology has contributed to the reduction of emissions from the first fan-jets in the early 1960s to the most recent marks of high by-pass fan where a reduction of 40 per cent in the weight of pollutants per unit weight of fuel burnt has been achieved. Improvements are likely to continue but the high combustion efficiency already achieved by the new generation of jet transport engines makes it increasingly difficult to forecast any major future advance in the pollution character-istics during flight. Important opportunities exist, however, for greater control on the airfield in the taxiing and idling conditions. Research and development is now being con-centrated in this area.

Pollution from airport transport

Because the petrol-engine-powered motor vehicle is in far more intensive use than the gas-turbine-driven transport air-craft, the motor vehicle would be expected to be a more serious factor in overall atmospheric pollution than the jet aircraft.

* Colloidal particles in a gas medium.

However, the petrol engine also operates at a much lower combustion efficiency, and the pollutant outputs per unit of fuel consumed are therefore much higher. Noxious exhaust products are at least ten times greater per pound of fuel consumed by the petrol engine, in the average cruising flight or road operating conditions. The problem around the airport is intensified through the concentration of aircraft and motor vehicles in rather closely defined areas, and the very low combustion efficiency of aircraft in the low-power taxiing phase of operations.

Motor vehicles (generally of above average power) concentrate on the air side of the terminals, on the aprons and in the vicinity of the operational buildings; on the land side they will concentrate in the approach roads, car parks, terminal buildings and cargo areas.

Improved standards of motor engine design now being introduced in most industrial countries will make an important contribution to the reduction of pollution from this source. Many urban areas have already achieved notable improvements, and the reduction especially in the concentrations of sulphur and nitrogen oxides could be read across into improvements in airport pollution from heating and industrial plant on the airport and in the peripheral zones.

Other aspects of pollution

We have not discussed the wider aspects of atmospheric pollution, such as that generated by aircraft in the upper air in the course of cruising flight. This phase of operations is totally remote from the airport environment, and the jet efflux will contaminate the upper air to a negligible extent, and in a manner likely to leave minimal effects for later generations. Special concern has been expressed over the impact of supersonic operations at altitude. ICAO has confirmed, however, that there is no scientific data to support the view that supersonic transport engine emissions at high altitudes will materially affect the weather or otherwise have a harmful effect on the atmosphere and human health.

There is little reason at all to doubt that the airline jet fleet contribution to overall atmospheric pollution is today negligible and is likely to remain small throughout this century.

4 Air Transport Operations: Changing Patterns and the Contribution of Technological Development

'That perfection of invention . . . which the human eye will follow with effortless delight . . . had in the beginning been hidden by nature and in the end been found by the engineer.'
—Antoine de Saint-Exupéry

'Knowing what it wants and realising how far aviation can serve these purposes, society can set the aims.'—D. Küchemann

In this chapter it has been thought most rational to bring together the airline operators' problems and those of the aircraft manufacturer and research worker in the airport operations field. Linked by a common concern to develop economic and effective air vehicles, all are interested in the community very directly in practice, but may appear not to be so by the community itself. This may be because some spokesmen for the public interest are in general unsympathetic to the objectives of the aircraft industry and to those of the air operators themselves.

Both sides of the industry have certainly had their problems – the uncertainties and vagaries of government support, the soaring inflation over the last decade and the restriction of fuel supplies. Its inherent vitality has kept it in business, as well as its value to the nation in peace and its significance in defence. The airlines, or, in other words, the operating side of the industry, have demonstrated one of the most remarkable growth records in industrial history.

The Press makes us all aware of the widening acceptance of civil aviation as a normal mode of transport, many being initiated by the availability of air tourism which has been growing by an average of over 15 per cent for the last five years, To woo the local public it would at the very least be inadvisable to offend the natural catchment which is within a short drive of their own airport. Within this area also will live most airline and airport staff. On a national scale no airline can afford to tarnish its image with the wider public by standing out as the operator of offensive aircraft, nor would it find sympathy with the air-tour promoting companies if it did so. Many airlines, moreover, are national carriers, and are more particularly obliged to respect the public interest.

Airline views on constraints in operation

The airlines in fact have expressed great concern about the deterioration of their public image caused by the increased noise nuisance of jet aircraft. In fact it has been found that they are not often able to do a great deal about it, but certainly since the mid-1950s a serious attempt has been made to undertake what was possible. As early as 1956, for example, BEA included in its outline requirements for a new type aircraft the statement that 'great importance is attached to the reduction of external noise'. Subsequently in 1963 a noise limit was set more precisely at 90 PNdB for an airbus project at a point four miles from the start of roll. Such guidelines have been instrumental in setting the new standards of noise which are being achieved by the trijets and airbuses in the mid-1970s within an internationally agreed framework.

While individual airlines have pressed for an improving external noise standard, IATA has worked more broadly for wider acceptance by the airline industry of its responsibilities to the community at large, and for concerted action by government research establishments and the aircraft industry to develop improved hardware. IATA has set up an Environmental Protection Advisory Committee made up of senior personnel drawn from member airlines and their own secretariat. Action in the technical, financial, legal and public relations field has been put in hand.

Of principal concern to airlines are the steps being taken to combat noise by the national aviation and airport authorities, some of whose methods have been stated to be detrimental to

the airline operator, and all of which add some form of burden
upon him. Perhaps the action subject to least dispute is the
steps still in hand to reduce the visible and invisible emissions
of gas and smoke. For one widely-used aircraft type a special
engine modification kit, comprising essentially a new form of
burner, entails a cost to the airline of approximately £3000 per
engine. An extensive programme of smoke control has been
put in hand which should be virtually complete by 1975.
The burden of cost, though partially met by US government
R & D funds, is in effect largely passed on to the airline industry.

As we saw in the last chapter, no major polluting effect is
incurred in jet aircraft operation since the gas-turbine engine
exhaust is signally free of the lead and sulphur constituents
which are principal elements in the air pollution caused by
earlier aircraft piston engines and by all automobiles.

Of greater current concern to the airlines is the noise abatement procedures in use at an increasing number of major
airports, as and when the noise levels become acute. While
striving to keep noise down to acceptable limits, airlines and
the pilots associations have emphasised the economic, operational and safety margins, which have in some instances been
reduced to a minimum.

The airlines would be expected to prefer the emphasis to be
placed upon the need to quieten aircraft and engines at source,
and for planning authorities to strengthen the control over
residential development and other land-use in the vicinity of
airports which has appeared to them to be far from adequate.
In this respect airlines look to governments and to the aircraft
industry to help them operate with minimum economic
constraint, and to seek these alternative and less restrictive
methods for combatting the noise problem in the airport
environment.

The airlines, through IATA, have called for international
standards to be set up with regard to aircraft pollution and
noise, emphasising that they await the improvement of noise
abatement technology, since at present the airlines and the
airports, key elements to the operating economy of a region, are
working at a disadvantage, meeting much of the costs of abatement and receiving most of the opprobrium.

IATA in a recent publication quoted an example from the USA:

> In recent years there was an agitation to stop jet aircraft flying into Midway Airport in Chicago. The agitation was successful, and jets ceased to use the airport; in a short time the economy of the area surrounding Midway became depressed, so depressed in fact that another pressure group rose up and demanded that jet transport aircraft be routed into Midway and this group, too, has succeeded. Jets are once again flying into the airport but the pressure continues – for an increase in the number of flights because it has been realised that they will revitalise the local economy (Ref. 24).

We have already discussed the use of night jet bans or curfews in Chapter 3. These are, in the airline view, ineffective and damaging to international air operations, increasing airline costs, extending the elapsed time of long-haul flights and greatly decreasing the value of air transport services to international commerce. We have already emphasised the local value of night flying bans, but the wider impact on long-distance flights is not often considered by an individual airport authority unless the facts are put fairly and squarely by the carriers.

Clearly the significance of these facts will depend upon the type of operators who use the airport. At major international terminals like London, Rome, Cairo, Teheran, Bangkok and Tokyo the interference with flight schedules could be extremely serious if night bans were universally adopted. The trend towards a wider use of curfews, which are a convenient and politically useful form of noise abatement, will continue to be watched anxiously by the airlines.

Jet-engine noise reduction

The major contribution of the airlines in the short term would be the modification of their existing aircraft to introduce quieter engines and components. Many of the world's major operators have been involved with the design, development and flight test programmes aimed at improvement in the noise level of engines in their fleets. We discuss below the principal forms of the so-called 'hush kits' which have been proposed for British and American engines.

A critical problem, however, is cost. So far as the British operators are concerned, an estimate of the total cost for modifying aircraft on the British register has been set at £100 million which when written off over about five years would comprise less than 1 per cent of total operating costs. Very exaggerated figures for the development cost of such programmes have also been quoted. Bearing in mind that it is likely to be some ten years before the new generation of quiet aircraft comprise more than 50 per cent of the transport movements at capital city airports such as Heathrow and Orly, it seems very well worth pursuing such projects with all the speed and energy which can be mustered. Unfortunately the UK has been slow in moving in the field of engine retrofit. British aircraft have a poor record for noise in the airport environment and the investment cost per individual British-built aircraft would be high in the initial stages.

One of the critical difficulties at Heathrow has been the fact that three of the noisier aircraft, all operated by the home-based operator British Airways, the old BOAC and BEA, viz. VC 10, Trident and BAC 111, comprise about 38·5 per cent of all transport aircraft movements at the airport (1972–73). For the significant British aircraft the development of a production package is still at an early stage.

Spey noise reduction programme

Results so far suggest that the optimum kit for the Rolls-Royce Spey engine in its various marks would include intake, bypass and tailpipe acoustic linings and a new design of the jet exhaust mixing nozzle. Comparable equipment is likely to be required for BAC 111 and Trident. Such kits should enable the noise levels to be reduced by 5 to 7 EPNdB and hopefully enable both classes of aircraft to meet the UK and USA noise certification standards. The full kit for the BAC 111 is likely to weigh at least 500 lb, to exact a 3 per cent reduction in take-off thrust and a 5 per cent increase in specific fuel consumption. The cost of this equipment has been reported in the *Financial Times* as being in the order of £70,000 per Spey engine, or £210,000 for a Trident and £140,000 for a BAC 111 without spare engines. At the very earliest such equipment would not be available for service before the summer of 1976.

Trident 'hush kits'

Work undertaken by Rolls-Royce and Hawker-Siddeley have led to proposals which could also offer considerable noise reductions on the Trident. Concentrating on the Trident 3B, the most recent mark, of which BEA now operates twenty-six, and which is equipped with three Spey RB 163–25 engines and a take-off booster engine, the RB 162–86, the programme offers a scheme for intake, bypass duct and jet pipe linings together with a six-chute nozzle for each propulsion engine. It has been stated by the technical director of Hawker Siddeley (Ref. 25) that the Trident 3B programme could be completed within three years of initiation at a cost of £4 million (1972 prices), while the whole Trident programme could be completed within four years at a total cost of less than £7 million. This overall cost per Trident is little different from that quoted for the BAC 111, but in practice and with inflation could be considerably higher than this. The impact upon Heathrow noise, if the Trident and BAC 111 noise programmes could be completed by 1978–79 would be immense.

The urgent need at the time of writing is for completion of the extensive flight test programme for noise evaluation on Trident and BAC 111 aircraft which comprised in 1972–73 over 35 per cent of total transport movements at Heathrow.

USA jet-engine noise programmes

Corresponding equipment for existing USA aircraft and engines are in a more advanced state. Early work on the Boeing 727 (Pratt & Whitney JT 8D engines) was aimed at reducing the fan-generated noise levels by use of sound-absorbent linings. This programme was pushed through without awaiting results of the time-consuming and more complex engine noise tests. Thus the manufacturer was able to meet US noise requirements (FAR, Part 36) as specified for new type designs. This retrofit cost was quoted by the Boeing company as $175,000 per aircraft at a weight penalty of 377 lb. Boeing report that since mid-1971 the majority of airlines have ordered the low noise version of the JT 8D at the higher price. This indicates a responsible attitude on the part of the airlines to questions of noise.

A co-funded FAA-Boeing programme has been energetically pursued to get further reduced noise levels in the approach, flyover and side-line cases with a minimum penalty in weight and performance. Parallel programmes are in hand for the Pratt & Whitney JT 3D engines (an earlier design than the Pratt & Whitney JT 8D) for the Boeing 707. Costs for this aircraft are estimated at £450,000 or $1,000,000.

Refanning the older jet engines

An alternative method of reducing older-engine noise is to increase the bypass ratio of the engines installed as in the case of the next generation of transports. With both Pratt & Whitney and Rolls-Royce fan engines of earlier design it is possible to replace the two-stage fan now in use by new designs of single fan of larger diameter. This will reduce jet velocity and hence noise in the jet stream. Lower specific fuel consumption and improved take-off thrust can also be achieved, and a reduced penalty in overall aircraft performance should be possible. Such development programmes are potentially very attractive to the airline operator, and the major carriers are today co-operating closely with their governments (particularly in the US and UK) and with the aircraft industry in sponsoring this work.

Considerable uncertainty has existed in the US as in Europe as to the economic feasibility of the refanning programme which could amount to hundreds of millions of dollars. The chairman of the US Civil Aeronautics Board, representing general air operating and regulatory opinion, has been unenthusiastic about the expenditure, and the budget for the financial year 1974 was cut down to $18 million (£8 million). However, the Air Transport Association was also concerned since, in effect, the limited budget has ruled out the refanning development work on the JT 3D engines of the Boeing 707 and the Douglas DC 8.

The world markets and the existing operating patterns of the US and British engines are not essentially different. The need to improve yet further the noise levels of the Boeing 727, Boeing 737 and McDonnell-Douglas DC 9, still selling strongly throughout the world, must be of key significance to the manufacturers and to the FAA. But the British aircraft also have important markets overseas and the noise problem in London, Manchester, Glasgow and in the capital cities of Europe requires further urgent action for the pursuit of the modification programme.

It hardly needs to be said that the airlines in the present climate of environmental opinion will accept a reasonable degree of performance deterioration and slightly higher operating cost when the 'hush-kit' equipment is available, and if it can be shown to give an important improvement to the levels of airport noise.

Noise abatement procedures

The control of the flight profile remains as the area of action most directly under the supervision of an airline. By profile we mean the aircraft flight path in climb and descent, from and on to the runway, its vertical trajectory and its azimuthal projection in plan. All airlines have readily co-operated so as to work the laid-down procedures and route patterns which generate the minimum noise disturbance, so long as they have met the minimum safety requirements, and conflict to the least extent with the need to operate an economic flight plan.

However, most airlines have very considerable reservations as to the desirability of operating to the noise abatement procedures now ruling at many airports. The control of noise for example at Heathrow by means of the monitoring points at a number of locations approximately where the built-up areas begin, has never been gladly accepted. BEA, for example have stated on several occasions that take-off monitoring at a single point (for one take-off) is not in the best interest of noise abatement, preferring the idea of monitoring the total noise exposure, hour by hour, on an area basis. It is, in the author's view, very undesirable that power reduction should be resorted to as a means of noise reduction for the benefit of close-in communities, who gain in many compensatory ways for their proximity to the airport, and at the expense of the more widely dispersed communities who suffer from the lower altitude of the aircraft which at some point needs to increase thrust (and noise) in order to climb away on to the airway.

Alternative techniques of high gradient take-off climb to maximise the use of high take-off thrust for a given operating weight have been employed to meet the needs of a particular airport. Pilots and airlines would prefer this, moreover, in the interests of standardising take-off and landing procedures. Another technique would involve acceleration up to a speed at which flaps may be retracted, allowing thereby a considerable reduction in thrust before the built-up areas and the point of noise monitoring. Noise is thereby reduced but at the price of low rates of climb until thrust is subsequently increased. This type of procedure when used at Heathrow would hit hardest those communities lying to the south-east where over the rising Surrey hills the increased thrust subsequently applied at the reduced altitudes (resulting from power cut) must raise noise levels higher than would have been the case with a straight through climb.

A better course of action now in general use for Heathrow take-offs is to climb out at a lower speed without early retraction of flaps and with a lesser reduction of thrust.

Slightly greater noise in the early climb phase is thereby exchanged for less noise over the communities further from the airport.

Approach and landing

In the approach and landing phases the airlines have had less opportunity to influence the trends of noise alleviation. It is the general practice to intercept the runway approach glide slope of 3° from below. Prior configuration and speed changes will have been completed, with their consequent noise increases, and the aircraft will be brought down at a steady approach speed from about 1500 ft altitude and eight miles out.

With existing aircraft little option is available for the improvements of this technique. While steeper approach angles are attractive in maintaining greater aircraft height in the early parts of the glide path, there is little hope of achieving high angles even in the first segment of the approach until improved handling characteristics are demonstrated in a wide range o adverse flying conditions for the jet aircraft types in extensive use.

The so-called 'two-segment approach' is now under active evaluation in the USA. Trials by American airlines using an area-navigation display seem not to have been entirely convincing, but results of ground measurement of noise under the flight path of the 727 aircraft at the outer marker, and at one mile from the threshold, gave a reduction of 18 EPNdB and 8 EPNdB respectively as compared with a standard instrument approach down a constant 3° glide path. Full development will require all of this decade, and may indeed need to await definitive STOL aircraft designs incorporating quite new concepts in low altitude handling, and offering steeper flight altitudes in all phases of the approach. We should note here that the pilot organisations such as IFALPA and BALPA are firmly against the hasty introduction of landing techniques employing the two-segment approach for jet transports of current design. We shall discuss the important airline contribution to these new areas of research and development when we consider later in this chapter the aircraft manufacturers' programmes of noise-defeating transport projects.

Introduction of new generation super-jets

The great hope for noise reduction at major airports in the future is the current programme for the introduction of new generation jets which was first initiated with the Boeing 747 in 1969. Two years later there followed the McDonnell-Douglas DC 10 and the Lockheed 1011 Tristar. These aircraft, which also have introduced the wide-body layout, have set the standard for a generation of medium- and long-haul jets which will without question dominate the airports of the world in the late 1970s and 1980s. Soon we shall encounter the transports from Europe and probably from Russia and Japan, first of which will be the European Airbus A 300 built by a consortium of manufacturers known as Airbus Industries, comprising Aerospatiale (France), Deutsche Airbus (West Germany and Holland) and Hawker-Siddeley (UK). These quiet, wide-body jet aircraft, already on order in quantity by major airlines, are gradually being introduced through the mid-1970s, so that at key hubs such as London (Heathrow) the operating aircraft population will be completely transformed through the 1980s.

In a recent study prepared for a public inquiry into air traffic over regions peripheral to Heathrow during the next decade, the following results were obtained. Note that by the year 1980 supersonic transport is seen as likely to be operating.

It must now be expected that these aircraft, because of their higher noise level (compared to new generation jets), would be required to operate from Maplin if the airport were to be built.

TABLE 4.1. *Estimate of Heathrow Jet Aircraft Movements in Thousands* (1980)

| Aircraft type | Passenger aircraft | | Cargo aircraft | Total jet transports |
	Short/Medium haul	Long haul	All routes	
New Quiet Jets	91	71	8	170
SST	–	7	–	7
Older Type Jets	131	7	51	189
Annual totals	222	85	59	366

A major step forward undoubtedly occurred with the introduction of national standards of noise level for transport aircraft following international meetings in Europe and the USA, and the acceptance by ICAO of the basic requirements and the principles of implementation. The UK and US requirements were discussed in Chapter 2 of this book.

In essence it is required that transport jet aircraft (excluding the SSTs) applying for certification after 1 December 1969 should demonstrate compliance with a level of noise to be measured in EPNdB (Chapter 5) at three test points, (*a*) on take-off at a point parallel to the runway, (*b*) on take-off at a point on the extended runway centre line, and (*c*) on the approach to landing.

Rising noise control standards

In addition to establishing a required level of noise in the USA for the major jet aircraft groups in defined conditions, the FAA made it clear that it has under study future control measures for 'other aircraft such as small turbine-propeller powered airplanes and reciprocating engined powered airplanes, vertical take-off and landing aircraft (including rotocraft), short take-off and landing aircraft and supersonic aircraft'. Such action, however, is in no way likely to be imminent in the USA or in Europe, since data on which to base decisions are still woefully lacking. Wider controls will, however, very certainly be implemented within the next ten years. Nor should we presume that the defined levels, as set out in the current requirements, will long remain fixed. The FAA has already in fact prescribed a noise 'floor' of 80 EPdB for all new aircraft in the given flight situations. This objective, it is stated, should be approached whenever 'economically reasonable, technologically practicable, and appropriate to the particular type design'.

It has been suggested that any airliner not meeting FAR Part 36 requirements by 1978 may be heavily restricted in domestic US operations and might even be grounded. Without much delay the aircraft industry achieved the levels called for in the various national requirements. The Boeing 747, the first major type to seek compliance, was rapidly in a position to meet the standard. Subsequently the DC 10 and the L.1011 proved compliance. There is every likelihood that new types of aircraft will exceed requirements by fairly comfortable margins, and we have already shown how the early 727 models have been modified so as to meet the requirements though not legally obliged to do so because of their early date of design and certification.

Engine development and the noise footprint

The three major engine designs which power the wide-body jets are the Rolls-Royce RB 211, the General Electric CF–6 and the Pratt & Whitney JT 9D. These are all single-fan engines with a bypass air ratio of about 5·1 with acoustically lined ducts and common design features. Significant was the elimination of inlet guide vanes which experience had shown to be a serious source of noise. Intensive research is in hand in the USA and the UK to gain more precise understanding of the basic physics of the sources of noise, and at the present time ground is being won largely by the gradual process of detailed analysis, testing and pragmatic advance.

However, the progress made is impressive and this can be demonstrated most clearly by the aircraft footprint and its

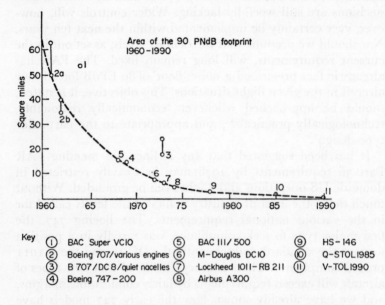

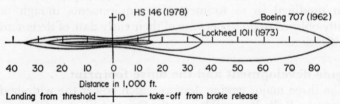

Key
① BAC Super VC10   ⑤ BAC 111/500   ⑨ HS-146
② Boeing 707/various engines   ⑥ M-Douglas DC10   ⑩ Q-STOL 1985
③ B 707/DC 8/quiet nacelles   ⑦ Lockheed 1011-RB 211   ⑪ V-TOL 1990
④ Boeing 747-200   ⑧ Airbus A300

Fig. 4.1   Aircraft Noise Footprints

remarkable shrinkage over the last five years. The footprint may be defined as the contour surrounding the runway which delineates a given level of noise from an individual aircraft during a take-off and landing. Thus in Fig. 4.1 the footprints shown for three typical aircraft of representative type indicate the boundary of the 90 PNdB noise level for a take-off and a landing in each case.

In its limit the footprint area which even today is in excess of 45 square miles for long-haul four-fan-jet transports could be reduced to little more than one square mile for a short take-off and landing aircraft (STOL) in the mid-1980s and to something less than the area of the airport itself in the case of a vertical take-off and landing aircraft (VTOL) in the 1990s. The footprint should not be confused with the NNI contours which have been discussed in Chapters 2 and 3. See also a more detailed discussion of NNI in Chapter 5. We here are excluding consideration of the frequency of movement, and view only the level of noise for one event sequence and the dispersion of that noise beyond the source of noise itself.

Sources of aircraft and engine noise

The principal sources of aircraft noise arise naturally enough within the engines themselves and in their exhaust systems. As aircraft noise levels become depressed, however, the basic level of the aerodynamic noise will probably never be entirely suppressed into the background. Work by General Electric has in fact shown that noise generated by the airframe generates in the sideline case a PNdB level only 10 dB below the certification requirement (FAR 36).

The work of Sir James Lighthill identified the objectionable characteristics of jet noise which have been the dominant factors in jet transport aircraft. He showed mathematically that when hot gas streams are ejected at high speed into the atmosphere a noisy mixing process occurs at the periphery of the jet which falls in frequency as it passes down stream.

The noise is approximately proportional to the efflux velocity raised to the eighth power. The human ear is sensitive to a varying extent to these sound waves, being most affected at about 3000 to 4000 cycles per second.

It is this variability in the human reaction which is the reason for adoption of the PNdB as a unit of noise measurement when we are concerned with the human reaction to aircraft noise (Chapter 6). Thus the best way to reduce the noise produced in the jet stream is to reduce its velocity. If we are to maintain engine thrust this can only be achieved by increasing the air mass flow. Thus has arisen the high bypass fan engine which generates high thrust by the use of a fan propelling rearwards a larger mass of air and with lower inner jet velocities.

The propeller-driving engine is in fact using a similar basic principle with an accompanying lower level of external noise.

Research has shown that jet noise can also be substantially reduced by improving the rate of mixing of the gases in the regions aft of the nozzles. Following work in the UK and USA a variety of devices have been evolved. These include corrugated and multiple nozzles, one form of ejector nozzle being so designed as to provide a noise shielding effect.

Fan design has created its own problems and in the RB 211 engine, which is the quietest engine in the world today for a given thrust output, noise from the fan has been reduced by the elimination of inlet-guide vanes, a source of fluctuating air forces and vibration in earlier fan engines which could be suppressed only partially by suitable acoustic linings.

Thus the reduction of jet efflux noise drew attention to the other sources of engine noise, and an extensive programme of research and development was directed towards identification of the other principal areas of noise generation.

Noise is also generated upstream of the final nozzle, and absorptive treatment has been successfully applied there. Intake and bypass noise can also be suppressed to a considerable extent by acoustic lining materials. Much work is in progress to determine the optimum lining materials for maximum noise suppression in relation to their weight. The weight problem is particularly critical, and even on the RB 211 the manufacturers have claimed that a further five decibels of noise reduction could be achieved if the weight increases could be justified.

The technology is in no way yet in a final stage, and each engine component is under active development for the cost-effective advancement of its engineering and noise performance. There is no reason at all to doubt that the noise signature of the new engines will show remarkable improvements within the decade.

The noise performance of current transports

In the next chapter we shall discuss in more detail the physical problem of noise and, for readers who wish to pursue the subject, aspects of the measurement of noise which because of the nature of our auditory senses is most conveniently recorded on a logarithmic scale. Thus the improvements achieved by the new generation of transports, which are advancing quite dramatically beyond the current standards of noise certification, can be easily underestimated by a brief study of the changes in the PNdB level. A decrease of 10 units in the PNdB scale represents a halving of the noise received and a 20-unit decrease represents a reduction to one quarter of the original value. Put another way, the flying noise of the L.1011 Tristar is approximately 20 PNdB lower than the Boeing 707 so that it would require 100 1011 aircraft to take-off simultaneously to equal the noise level of one 707 taking off. It is no exaggeration to say therefore that the improvement to be expected by the new generation of aircraft, year by year, will be dramatic, and the widespread underestimation of the significance of these facts is to be deplored.

Table 4.2 provides a summary of the principal transport aircraft in operation in 1974 with their maximum weight and runway requirements. This list gives some indication of the types in operation and their airport demands.

In the table also we have shown the noise levels generated by these aircraft at their maximum operating weight as measured (or in some cases estimated) at the noise certification points defined in the US and UK regulations. These regulations were defined in Chapter 3. Some difficulty arises from the differences between the PNdB and the EPNdB units. This will be clarified in Chapter 5. Here we must content ourselves by saying that the PNdB, used in the NNI computations and in the footprint calculations used for Fig. 4.1 are not to be directly compared with the more precise EPNdB units which are employed in noise certification for the three critical cases, viz.:

A. Flyover noise levels.   B. Side-line noise levels.
C. Approach noise levels.

TABLE 4.2. *Summarised Airport Specification for Transport Aircraft*

	Type	Max Pass.	Capacity Cargo* lb	Gross Weight lb	Landing Weight lb
U.S. Aircraft					
Boeing					
707	320 B	160–189	28,200	336,000	247,000
727	200	120–189	21,000	160,000	154,500
737	200	115–130	13,135	116,000	103,000
747	200 B	375–500	103,170	778,000	564,000
Lockheed					
L. 1011	Tristar	256–400	45,750	430,000	358,000
McDonnell-Douglas					
DC 8	50	140–176	20,850	315,000	207,000
DC 8	61	230–259	66,665	325,000	240,000
DC 8	30	100–115	26,534	98,000	93,000
DC 10	30	250–380	46,000	555,000	403,000
UK Aircraft					
BAC					
111	500	100–119	5,000	104,500	87,000
VC 10	Super	150–163	20,000	337,000	237,000
Concorde	Anglo-French	108–128	7,000	385,800	240,000
Hawker-Siddeley					
Trident	3B	160–179	11,000	158,000	128,500
748	2A	40–62	3,300	44,500	43,000
Other European Aircraft					
Aérospatiale					
Caravelle	10 B	64–99	5,700	114,650	109,130
Fokker					
F.28	2000	48–79	4,700	65,000	59,000

* Cargo Capacity is approximate only.
Data Source: ASA Information Office, Aircraft Manufacturers *Janes all the World's Aircraft*.

Power Plant Number, Make, Type	Take-off Field Length ft	Land-ing Field Length ft	Noise Levels in EPNdB at Certification Points		
			Flyover	Sideline	Approach
4 × P & W JT 3D	10,020	6,250	114	108	120
3 × P & W JT 8D	8,050	4,150	101	100·5	109·5
2 × P & W JT 8D	4,700	3,750	98	101	112
4 × P & W JT 9D	10,500	6,200	111	101	112
3 × RR RB211	7,950	5,680	98	95	103
4 × P & W JT 3D	9,450	5,400	115	106	117
4 × P & W JT 3D	9,980	6,140	117	103	117
2 × P & W JT 8D	9,200	5,100	95	103	109
3 × GE CF6	10,500	5,330	104	97	107
2 × RR Spey 512	7,300	4,720	103	108·5	102·5
4 × RR Conway 550	8,300	7,000	110	113·5	115·0
4 × RR Olympus 593	10,950	8,880	114	111·0	115·0
3 × RR Spey 512	7,500	5,920	104	108	112
2 × RR Dart RDa7	3,070	1,980	92·5	96·3	103·8
2 × P & W JT 8D	6,850	5,180	99	102	107
2 × RR Spey 555	5,500	3,540	90	99·5	103·8

A number of important conclusions can be drawn from these figures. The remarkable drop in EPNdB levels for new aircraft types in all three cases is self-evident. The need to improve the large number of older jet aircraft in service is also apparent.

Table 4.1 has indicated the number of these likely to be in operation in 1980. We have shown there that in the case of London Airport (Heathrow) even in 1970, 54 per cent of transport movements are likely still to be with older jet aircraft and supersonic aircraft.

The USA Aviation Advisory Commission reporting in early 1973 on the long-term needs of aviation, indicated a far greater contribution from jet-engine nacelle treatment for the older aircraft (primarily on the Pratt & Whitney JT 3D & 8D engines) than from improved operating procedures or new quiet engines for the next eight to ten years. After that time, in the early 1980s, a notable contribution is expected from nacelle treatment and the re-fanned engine programme.

The continuing trend towards larger aircraft, which thereby reduces the rate of growth of aircraft movement, must be introduced as a favourable factor, since the overall noise produced per passenger carried is being steadily reduced by these developments.

Two-segment approach

The introduction of two-segment approach techniques must be expected within the next ten years, because it will inherently allow aircraft in the landing phase, when it is most difficult to achieve great improvements by conventional methods of noise reduction, to be kept at a greater altitude from the ground. The method used in the recent trials conducted by United Air Lines in April 1973, with Boeing 727 aircraft on schedules between San Francisco and Los Angeles, was to approach at 6° up to $1\frac{1}{2}$ miles from the threshold, and then to take up the normal glide slope of 3°. We have commented above on the cautious attitude adopted by airline technical departments and by the airline pilot's organisations. It seems likely that with the newer generation of medium-range aircraft, with reduced take-off and landing distance requirements, it will be possible for an airline to introduce the approach aids and avionic equipment which will satisfy technical opinion. Moreover, the performance and handling characteristics inherent in such designs will more readily provide acceptable margins to satisfy airline project engineers and flight-deck crew alike. Problems are not all solved, however, with the improved field performance since the increased climb and descent gradients will be acquired by inevitable design compromises, including high thrust weight ratios and a wider cabin-floor gradient range.

The new era of air transport noise

Fig. 4.2 shows a symbolic representation of the airport noise problem in the airport environment. Using the area of the 90 PNdB contour, it shows in curve A the steep increase in the area affected when the jet transports came into service in the early 1960s.

Airports were developed to meet the increasing requirements of the jets for extended runways, and the increased area of the new airports built during the 1950s and 1960s, even up to the present time, is shown in the lower curves B, where it is indicated that superficial airport areas have grown from about 4·5 sq. miles to about 35 sq. miles in the twenty-five years 1950 to 1975.

The full line curves show the first introduction dates of the new aircraft with their increasing PNdB areas, and also the initiation of the new airports, but clearly both new aircraft and airports remain in extended use, the aircraft for at least fifteen years and the airports for at least twenty-five years, dependent upon obsolescence and development. Thus the curves have been extended to the right to indicate the period of use, and it is seen that by the mid-1970s, when it is likely that peak values for the size of new airports will have been reached, the impact of the reduced PNdB areas for the new aircraft (reflecting – *both* reduced noise and take-off requirements) will reduce the *land* areas likely to be seriously affected by noise outside the airports themselves.

The situation in the late 1980s is likely to have reached a new stability where the high noise areas of aircraft operations could be comparable in extent to the airport areas required to meet the take-off and landing requirements. This is the situation with which we were familiar in the early 1950s, except that today other air traffic and airport design needs will dictate the layout

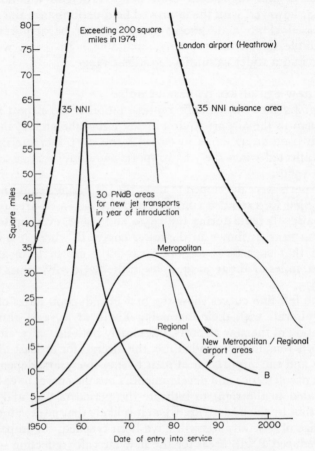

Fig. 4.2   Airports and Noise Impact Areas

and the areas required. However, because the PNdB footprint is of a different shape generally to that of the airport areas required for landing, take-off and general airport operations, it will not be easy to retain the high PNdB contours entirely within the periphery of the airport. This will be closely approached in the STOL era of the 1980s, but it is unlikely to be fully achieved until VTOL operations are more generally in vogue, which is unlikely to be before the 1990s.

New aircraft type developments

New aircraft types now under active design or development have naturally incorporated important advances in noise alleviation. Arising out of the government research programme and intensive R & D in the aero-engine industry, noise levels for new transports have now been set well below the current requirements. A summary of some of the key designs and their estimated noise levels may be given.

The Boeing Company released details in 1973 of 2-, 3-, and 4-engine design studies for the so-called 7×7 transport with a new mark of Pratt & Whitney JT 10D engine. Work was geared to likely 1977–78 airline deliveries, and thus an important contribution to fleets in replacement of current 2-, 3-, and 4-engined types might still be anticipated by the early 1980s even when taking the fullest account of the fuel crisis. Design estimates of the 4-engined version suggest noise levels well below the US regulations of FAR 36 – reductions of 13 EPNdB in the side-line case, and 7 EPNdB in the flyover case have been mentioned by Boeing project engineers.

A 300

The European Airbus now in development in France, Germany, Holland and the UK could with Pratt & Whitney engines designed within the current state of the art also generate average PNdB levels below that of the Lockheed Tristar at the critical certification measuring points. First tests have shown that the certification levels can be improved upon without difficulty, and 8–10 dB below requirements is the objective.

Europlane

Europlane, a twin shoulder-engined project sponsored by the manufacturers – BAC, SAAB-Scania,* MBB† and CASA‡, was estimated to generate noise levels 10 dB lower even than the current certification requirements. Rolls-Royce RB 211 or General Electric CF 6 engines were to be used. The 90 EPNdB noise contour was estimated to cover less than two square miles, compared with the 15–20 square miles of most existing twin jets before 'hush-kit' modification. Thus, in a precisely analysed case on Runway 28 Right at London (Heathrow), when using conventional 3° glide slope and touching down inside the threshold, it is estimated that the 90 EPNdB contour lies entirely within the airport boundary. This should be typical of new generation aircraft described as 'Reduced Take-off and Landing Aircraft' which will require less than half the available runway at a major airport like Heathrow. In this design also critical points on the noise carpets should be more than 10 dB below the existing certification requirements. This project has now been cancelled, but is quoted as typical of the state of the art in 1973.

H.S. 146

This project, which received UK Government backing in 1973, may be considered as typical of the fourth generation of jet aircraft designed specifically for the medium-short-haul market, and making maximum use of the latest quiet engine technology as well as providing an early example of a 100-seat transport with reduced take-off landing and field requirements. Certainly in the UK and European airfields this aircraft will be creating an important impact in the late 1970s. It may most simply be conceived as a Viscount replacement, and its noise level should be little greater than the latter. Fig. 4.1 shows the contour for the 90 PNdB noise levels as indicated by the manufacturer's estimates. First flight is likely in 1975. Airline service is scheduled for 1976.

* Sweden. † West Germany. ‡ Spain.

Supersonic transport aircraft

The supersonic transports, Concorde and the US projects, were originally designed to meet the airport noise level of the first 4-engine jets in the early 1960s. The world-wide concern with noise, which led to the promulgation of regulations for the control of noise at airports and in the design stage, came upon them when at a late phase in their development. The FAA and the UK regulations in fact specially exclude the SST from the necessity to meet the full noise requirements. Certainly the Concorde (fitted with the Olympus 593 MK 601 engines), with an improved exhaust-silencing system will be of comparable noisiness to the currently operating 4-engined jets on flyover and approach even though the lateral noise will be about 4 PNdB higher than that of the current fleets. Work has been in hand to produce the most effective retractable silencing system for production aircraft. In this system, known as the 'spade' silencer, a series of retractable spade-like devices are deflected into the jet efflux to divert some part of the high velocity exhaust gas into the annular space between the primary and secondary flows. Some loss of thrust is thereby incurred, but this is acceptable in a retractable system which will have a negligible effect in cruising flight.

In the full take-off power case – when lateral noise is critical for the SST – estimates have shown that the spade silencers should show a 4.5 PNdB improvement for a loss of thrust of 5 per cent. This should reduce lateral noise to 114 PNdB which is still above the Boeing 707–320 and the MacDonell-Douglas DC 8–60 series aircraft. Further noise reductions are anticipated from later exhaust nozzle designs, including the TRA design, the buckets of which are used for thrust reversal and vary the final nozzle area.

Even were a new SST project to be conceived today it is difficult to see how a major breakthrough in airport noise levels could be achieved when the critical cruise requirement for power plant design are a high thrust/drag ratio, implying high jet velocity and a straight-through turbine cycle. The Director General of SNECMA, M. Garnier, has noted that 'if the principle of a straight flow cycle is maintained a significant gain in s.f.c. during supersonic cruise and in subsonic flight could be expected by raising the temperature from 1100° to 1300°, and the pressure ratio from 14:1 to 23:1. . . . I can only confirm today the validity of the decision made by Rolls Royce and SNECMA as early as 1961 to adopt a straight-flow cycle for the Concorde's engine.'

The problem of airport noise nevertheless remains, and in the first years of Concorde service, the new generation of wide-body jets with greater operating weights will be flying from the same civil airfields as Concorde and with a much more acceptable level of noise.

Noise levels estimated for the Concorde in comparison with other older jet aircraft are shown in the following table. Concorde figures are with the improved 'spade' silencers and specified in EPNdB.

These figures were issued in 1971 by Aerospatiale with reference to the US FAR (Part 36) regulations applying to aircraft for which application for certification was presented after 1967. Thus the Concorde is not required to meet the rules, but will naturally be operating in an environment increasingly dominated by turbo-fan engines with high pass ratios.

TABLE 4.3. *Comparative Community Noise (EPNdB)*

	Boeing 707-320 B	DC 8–50	Concorde
Side-line 0·35 NM	108	106	111
Flyover 3·5 NM	114	115	114
Approach 1 NM	120	117	115
Total	342	338	340

A direct comparison with the Boeing 747 and the large Trijets is of course less favourable to the Concorde, the Trijets being able to achieve a 5 to 10 PNdB improvement in approach, and up to 15 PNdB improvement in take-off, compared with Concorde noise levels as estimated now for operators in the late 1970s.

It must be borne in mind that the Concorde is a first generation aircraft of revolutionary type. The true impact of supersonic transport will not be felt until second and third generation projects are launched, probably these will be of greater range and payload capacity, and almost certainly with modified engines of inherently better design and noise characteristics.

First impressions of the noise levels of the Soviet supersonic transport, the Tupolev TU 144, are that its noise level is higher than that of Concorde. There is no doubt that the engine and airframe industry in the USSR have so far taken the subject of airport noise very much less seriously than we have in the Western world. The likelihood of operation by international airlines, apart from those operated by states politically aligned with the Soviet Union, is small. The present views on noise levels at airports, and the internationally agreed airworthiness standards (ICAO Annexe 16 covers noise) are believed to be a major deterrent to international acceptance of the Soviet SST. Visits by the TU 144 must be expected, however, at most of the capital city airports in Europe and the Far East as from the mid-1970s.

There can be no doubt that a future generation of supersonic transports will be required to meet at least the current ICAO and FAR 36 noise standards, since by the early 1980s transport aircraft will have reached even lower levels of noise production, and states will not be likely to accept jet transports of any kind which need a special dispensation for operation through their major airports.

Thus will the aircraft industries in Europe and North America move into their stride to provide a new generation of transport aircraft of ever increasing capability, economy, technical excellence and quietness.

Noise shielding

Serious attempts are being made to ensure that the new aircraft will achieve greater shielding of engine noise from the ground by parts of their own structure than has been the practice in the past. Overwing engines have provided a certain amount of shielding, and the VFW 614, a West German design for a short-medium-range transport has been exceptional in this respect, but the Europlane project and the Boeing 7×7 designs have also shown a keen appreciation of the possibilities in this regard.

Noise reduction from this source of about 4 EPNdB has been estimated for Europlane. Forward noise shielding and rear-arc shielding should be distinguished, the former being of particular significance in the landing phase, where improvements upon the existing level of noise certification may prove very difficult to achieve. A very interesting field of design lies open here to the manufacturer, which may offer great opportunity for new concepts and original layouts in future transport aircraft. Other prospects for shielding of noise from the ground lie in the lifting vortices shed from aircraft wing tips. Aircraft designs can be developed to take advantage also of this effect, and this is now under study in the aerodynamic research institutions of many countries.

Government research and development

Fundamental to much of the aircraft and engine development discussed above are the research programmes sponsored by those major industrial states with an aeronautical industry of some size.

In the USA the Space Programme is still the prime spender of research funds in the Aeronautical and Space Administration, but massive increases were recorded in 1973 for aeronautics itself. In the NASA Aeronautical Research and Technology Budget for 1973 no less than $12 million were spent on experimental quiet engine programmes, and $27·5 million on quiet experimental STOL Aircraft. This excludes many other items of closely related engine research and development, and the Vertical Take-off and Landing (VTOL) programme, much of which is geared to aeronautical advance in the interests of the environment.

In the US industry and in the research establishments, further extensive programmes aimed at noise alleviation are in progress. That part of the work of Pratt & Whitney, General Electric, Boeing and McDonnell-Douglas aimed at the deeper understanding of noise, and the development of hardware for 'hush-kits' modification programme, and for noise orientated new design evolution is reported to run to $20–30 million per annum (Ref. 26).

Quiet STOL research in the USA

NASA is re-investigating its Quiet Experimental STOL research programme by participation in the US Air Force's

Advanced Medium STOL Transport prototype project (AMST) An earlier NASA Questol* project was cut recently by US government budget changes. The new reduced programme is part of a 'Quiet Propulsive-Lift Technology' project proposed to cost $2 million in the fiscal year 1974. NASA's Ames Research Centre at Moffett Field, California, will manage the NASA part of the project, which will define the research vehicle requirements in the first stage with analytical and wind-tunnel work on aerodynamic and propulsion systems for alternative turbo-fan propulsive-lift concepts.

NASA objectives are to some extent different from that of the US Air Force which has a primary aim to replace the Lockheed C-130 military transport, with a lesser requirement for high performance at moderate cost and low noise level which must remain the objective of the civil aeroplane.

Research and development in the UK

In the UK the research programme related to aircraft noise has been estimated in government papers to have exceeded £1½ million in the year 1972–73. This would include the work undertaken in the aircraft and engine industries, that undertaken by government establishments – the National Gas Turbine Establishment (NGTE) and the National Physical Laboratory (NPL), and work at universities such as Southampton and Loughborough and the College of Aeronautics, Cranfield. At the NGTE at Pyestock test facilities have recently been extended for engine jet and duct liner testing.

Such work continues apace; more rather than less will in future be spent annually in the search for reduced noise levels. It has become a matter of community protection and overseas sales success as well. In all respects communities in all parts of the world will be the winners.

The Bristol Division of Rolls-Royce (1971) Ltd and Dowty-Rotol are hoping to extend their ultra-quiet demonstrator engine/programme as a research project based on the Rolls-Royce SNECMA M 45 engine and the Dowty-Rotol variable pitch fan. This will include French participation in the demonstrator programme.

The study and testing of noise-reducing techniques is in

* Quiet aircraft with short take-off and landing.

active progress with both engine companies. The geared
variable pitch fan allows engine output to be maintained with
considerable noise reduction. The fan turbine can run at high
speed and high efficiency, and by means of gear reduction a
larger low-speed fan can provide high thrust with low noise
level. Variable-pitch fan blades also allow the fan to run at
optimum pitch and this, too, gives a reduction in noise. The
variable-pitch fan also offers improved handling qualities and
enhanced aircraft safety.

This is useful also in the baulked landing case and for thrust
reversal on the ground. Aircraft ground manoeuvring could be
more precise since reverse thrust could be used down to zero
aircraft speed. This will have added value in near-zero visibility
conditions after landing. A faster response rate is offered. This
is of increased importance for thrust control on steep approach
paths now being proposed for lower noise levels on the ground.
Even the optimum engine conditions for low noise level can be
selected. Acoustical lining in the engine ducts further reduces
the perceived noise.

Japan

The Science and Technology Agency of Japan has now also
decided to carry out a two-year programme to make research
on a short take-off and landing (STOL) transport system
linked to improved airport environmental conditions. It
recently recommended a co-ordinated plan of aircraft (STOL),
airport and air navigational facility development, noting that
Japan had lagged behind in the development of STOL aircraft. The work will be divided into the five parts, summarised
below, under the aegis of the Co-ordination Bureau of the
Science and Technology Agency.

1. Research on STOL transporatation estimation. By the
   Transportation Economy Research Centre.
2. Research on STOL airports. Airport consultants engaged.
3. Research on STOL System Technology, e.g. navigational
   facilities and flight technology. The Radio Research
   Laboratory of the Transport Ministry and the National
   Aerospace Laboratory.
4. Research on STOL aircraft. The Society of Japanese
   Aircraft Constructors.

5. Comprehensive Research on STOL transportation. A STOL System Study Group of the Co-ordination Bureau.

Canada

The National Aeronautical Establishment in Canada, under the National Research Council, has had a programme of acoustical research in hand since the mid-1960s when special test facilities were set up in co-operation with the High Speed Aerodynamics Laboratory. Work on fundamental research and on the practical development of advanced transport concepts is not confined to the major industrial states. The Canadian Government has expressed deep concern over airport developments which conflict with the environment. The STOL work in Canada is well known. It is based on the exceptional success of Canadian aircraft projects with high capability in the short take-off though generally of smaller size than is required on major trunk routes.

General aviation

Although the principal impact of smaller aircraft on the environment is usually felt through the jet executive aircraft this is by no means always true. A small airfield where light single-engined aircraft predominate can create a serious noise nuisance, unless the movements are properly controlled and circuits are determined in relation to the surrounding residential areas. Light club aircraft can generate a noise nuisance even though their PNdB is low – even below 80 units on the scale and where no impact is made on the Noise and Number Index (NNI).

A great improvement in the noise level of a number of twin piston-engined aircraft is desirable. Much of this noise is created by high propeller r.p.m. on take-off.

Training schools with a routine programme of take-off and landing circuits can create special problems. Business and executive type operations are generally trouble-free since movements are well dispersed, and the individual resident aircraft generally takes off once only per day and then early in the morning, returning the same evening. Jet executive aircraft are still a major problem and certain engines of early design such as the Viper 600 could not readily be suppressed without considerable cost and weight penalty. Even so, a hush-kit programme is being actively pursued. More recent executive jets such as the Cessna Citation (with Pratt & Whitney JT 15D engines) have on the other hand shown what can now be done. For this aircraft, capable of operating from 2500 ft airfields, a level of EPNdB has been demonstrated which is at least 15 points below the required FAR 36, UK and ICAO certification level, in all three of the test cases. Most new designs of executive jet have a similar inherent capability of conforming with the new community standards. The development work of the engine manufacturer should also be mentioned.

The TFE 731 turbo-fan of the Garrett Corporation has been designed to provide low levels of noise for executive and small jet transports. In typical installations on the Swearingen SA–28T, the Dassault Falcon 10, and the Lockheed Jetstar, the 90 PNdB footprint is estimated to be about three square miles in area. This of course would be well within existing jet transport standards.

Jet executives will surely therefore have become of negligible significance in the airport environment by the 1980s even though total movements will have increased considerably. But there is little doubt that a very much more extensive under-standing of the short-term problems already being posed by large-scale general aviation is now required. The growth in movements is likely to be continuous and to create noise nuisance for at least a decade. Here lies a problem area which must be seriously faced by governments and by the aircraft industry in the immediate future. In the longer term the prospects seem more assured.

5 Fundamental Aspects of Noise and the Sonic Boom

Colin Waters

'All the available evidence, both in this country and
abroad, indicates that aircraft make a minimal contribution
to total air pollution.'—Royal Commission on Environmental Pollution

'For tis the sport to have the engineer
Hoist with his own petard.'—Shakespeare (*Hamlet*)

The study of the problem of noise and pollution around airports
is based upon a number of scientific disciplines and principles.
In order that the subject may be better understood, this
chapter seeks to explain the principles of noise generation,
subjective response to noise pollution in some detail but without
the introduction of a mathematical treatment. Some readers
may prefer to study this chapter on a second reading whereas
others may prefer to seek amplification in more detailed texts.
We refer them particularly to Refs 14, 27, 33, 36 and 37 in this
respect.

Some general principles

Noise has often been defined as unwanted sound and this
definition immediately leads to one of the great problems in all
noise work, namely that what is unwanted sound to one person
is pure music to another. Noise can also be thought of as wasted
energy, and can be considered to be undesirable in the same
way as the heat from a loaded bearing or the exhaust products
from the internal combustion engine.

However, before considering the subjective and philosophical
definition of noise it would be as well to look closely at the
mechanisms of sound generation and propagation. Sound is
created by any vibrating body which, in turn, sets the air near

to it into vibrations. These vibrations are transmitted from one molecule of air to the next. If a receiver is placed in the path of these vibrations then the pressure vibrations are sensed. If the receiver is a microphone the air-pressure variations affect the diaphragm and are then transferred into variations of electrical current and can be used to affect a meter. If the receiver is the human ear then the vibrations affect the ear drum, and its movement is transmitted to the hearing cells in the inner ear, and thus to the brain so as to give the sensation of hearing.

The character of sound is dependent upon the speed at which the source is vibrating, and also upon the amount of movement and the size of the vibrator. The speed of vibration gives rise to the property of frequency or pitch of the sound, whereas the movement and size of the vibrator give rise to the property of amplitude or loudness of the sound.

As an example, a violin string of a certain length will vibrate at 262 cycles every second (or 262 Hertz) when bowed. This string will do this whether it is attached to a violin or merely supported between two firm points. Thus the pitch of the note is established. However, its amplitude is greatly increased when attached to a violin, such that the vibrations are transmitted via the resonant chamber of the instrument, so that the body of the instrument vibrates in sympathy with the strings. The size of the radiating vibrator is then changed from the string itself to the much bigger area of the violin with the consequence that a bigger volume of air is set into motion and a louder note is heard. The frequency remains at middle C but the amplitude and loudness is increased. Incidentally, considerations such as 'attack' and 'colouration' have only musical meanings, which are dependent upon the method of hearing and the quality of the instrument. If we now consider our violin being played in the middle of a very large field we can look at the way in which the sound is propagated and the various ways in which it can be attenuated (or reduced). The sound is radiated in all directions evenly, and can be thought of as a hemisphere of sound waves.

This rather idealistic treatment assumes, of course, that all sound radiated downwards is absorbed by the ground without reflection. Sound is radiated in such a way as to obey the 'inverse square law', which means in effect that the intensity of the received sound varies as the inverse of the distance from the source squared. Thus if the distance of the observer from the source is doubled, then the received intensity of the sound is reduced to one quarter. If the distance of the observer from the source is trebled then the received intensity of the sound is reduced to one ninth, and so on. Thus it can readily be seen that increasing the distance of the observer from a sound source rapidly reduces the intensity or level of the sound heard. There are other ways in which the received level could be controlled, for example by putting the player in a sound-proof box so that no noise could be radiated, or, conversely, putting the observer in a box so that no noise could be received!

This rather simple example gives the elements of noise control technique – the noise can be reduced at source, the distance of the observer from the source can be increased or the observer can be protected from the receipt of the noise. The choice of the particular method employed will depend greatly upon the economic constraints imposed.

Characteristics of aircraft noise

The noise produced by an aircraft is different from that produced by a bowed violin string in that the aircraft noise produces a spectrum of frequencies all of which have their own amplitude, whereas the violin string produces essentially one dominant frequency having its associated amplitude. If the noise from an aircraft is recorded it is possible electronically to break down the spectrum into its component parts and to look closely at each small band of frequencies. In this way it can be seen that the aircraft noise is made up of various components. The vibrations set up in the air by the action of the engine compressor blades, the turbulent mixing of the jet with the surrounding air, the sympathetic vibration of engine panels, can all be identified and their effect upon the total noise output of the aircraft assessed. Fig. 5.1 shows such an aircraft noise spectrum and identifies the component frequencies of the various vibrating bodies.

The hearing mechanism of the human ear is such that the brain is more sensitive to some frequencies than others. The frequencies which can be heard by the average adult range

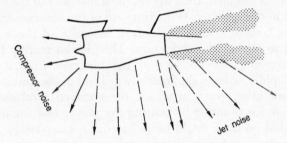

Mechanical and combustion noise

Fig. 5.1a   Components of Aircraft Engine Noise

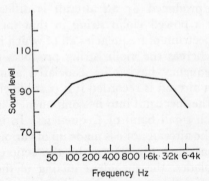

Fig. 5.1b   Octave Band Spectrum of Aircraft Take-off

from 20 Hz* to 20,000 Hz although this range varies according to individual physiological differences, and is reduced with age. The sensitivity characteristics are shown by an ability to hear high-frequency sounds at a lower level than low-frequency sounds. The ear is also capable of hearing sounds vastly different in level. The ratio of the loudest level of sound which can be heard to the level of sound which causes pain is 1 to 10,000,000, give or take the odd thousand. Dealing with this range causes no concern whatever to the ear, but it does put a strain on anyone trying to describe the level of noise reaching him. If one imagines the difficulty of trying to draw a graph with a base scale ranging from one to ten million one can get some idea of the practical difficulties. To overcome the problem, noise levels are described in terms of the comparison of one noise to another. In order to do this sensibly a reference level has been chosen which is the quietest audible sound.

Noise measurement

The methods of noise comparison is based upon logarithms and employs the unit of the decibel. To find the sound pressure level (SPL) of any sound it is compared with a reference sound pressure level of 0·0002 Pascals in the following way.

$$\text{Sound pressure level} = 10 \log \left(\frac{P}{P_{ref}} \right)^2 = 20 \log \cdot \frac{P}{P_{ref}}$$

Looking again at the vast scale of sound which can be heard we can see that the extremes of the scale are represented by one and ten million, thus the SPL of the threshold of pain is

$$\text{SPL} = 20 \log \frac{10,000,000}{1} = 140 \text{ decibels (dB)}$$

This, of course, is a much easier scale to handle. Some typical sound pressure levels are given below to help in locating various noises on the scale.

An instrument to measure noise can be made completely objective, and if, say, a noise of 100 dB at 500 Hz will give the same scale deflection as 100 dB at 5000 Hz, this instrument is

* Hertz – a measure of sound frequency.

have a linear or flat response to frequency. How-
~~~~ does not react in this way and, as was mentioned
~~~~ more sensitive to high-frequency sound than low-
~~~~quency sound. Thus to the ear the 100 dB at 5000 Hz would
appear louder than 100 dB at 500 Hz. SOUND PRESSURE
LEVES

Typical Sound Pressure Level dB	Noise Source
160	Service rifle at ear level
130	Jet aircraft taking-off at 150 metres
100	Inside a foundry
80	Ringing alarm clock at 1 metre
65	Busy general office
25	Very still day in the country away from traffic.

It therefore becomes obvious that a linear meter reading of
sound has only limited use when related to the expected
response of the ear. What is required is that the meter be made
subjective in such a way that when a sound which has a wide
band of frequencies in it is measured, the low frequencies are
made to give a lower than linear reading and the high frequencies are made to give a higher than linear reading. This
can in fact be easily achieved by building electronic weighting
networks into the circuits of the sound level meter so that the
reading on the scale more nearly corresponds to the response of
the ear. One such network is known as the 'A' weighted scale,
and when used for the measurement of a wide range of sounds
gives results expressed in dBA. We would then expect 100 dBA
of weighted low-frequency sound to be subjectively equivalent
to 100 dBA of weighted higher-frequency sound. There are
other weighting networks such as dBB, dBC and dBD which
are more nearly correct for more specialised sound sources.

Perceived noise levels

The consideration of a complex event such as the flyover of
an aircraft necessarily demands a complex method of measurement. It should be emphasised that as far as the ear is concerned there is no problem; the individual experiences the
event, and a set of reactions occurs which gives rise to his
recognition of an aircraft flyover of a certain loudness. The

problem arises when trying to ascribe a value to the noise which relates to the human experience. The use of a meter calibrated in dBA approximates quite well, but work originated by Kryter in America has resulted in the concept of the Perceived Noise Level, expressed in PNdB.

Put very simply, this method breaks the total noise into a series of individual frequency bands, and the noise level of each individual band is summed in a predetermined manner to give the overall noise level of the event. This noise level is then the Perceived Noise Level. As is to be expected, there is a relationship between the scales of measurement, and for all intents and purposes the PNdB level of the noise for a large jet aircraft is equal to the dBA level plus 12 units.

The requirement of the Aviation Authorities to issue certificates of airworthiness which include noise specifications has led to a sophistication of the basic perceived noise level to include such factors as duration of the event, and the pure tone content of the noise. These factors are included in the Effective Perceived Noise level, which is measured in EPNdB. As this noise level is associated with the certification of an aircraft there are very complicated rules laid down for its measurement. In Chapter 2 we have discussed the background to aircraft certification and the ICAO, FAA and UK government regulations pertaining to it.

Noise from the combined movement of aircraft: The NNI

It must be emphasised that the preceding methods of aircraft noise measurement are concerned only with single aircraft events, and it would be unwise to attempt to estimate the reaction of the community exposed to the noise from the operation of an airport by referring only to the dBA, PNdB and EPNdB reading of an isolated aircraft. What is required is some method of obtaining an overall impression of the amount of noise which is radiated over a period of time. The work of the Wilson Committee on Noise, which published its findings in 1963 in Cmnd 2056 (Ref. 27), investigated this aspect of airport noise assessment. They commissioned a social survey around London (Heathrow) Airport which probed by means of questionnaire interview the reactions of the population living in the area of the airport to the noise experienced. At the same time a physical noise measurement survey was carried out to assess the level of noise in the area. The survey data gave rise to 58 socio-psychological variables of community reaction, and the physical noise data defined 14 different parameters which could be picked out when characterising the noise climate. The relationships between these variables, or their correlation, were determined, and it was found that the significant physical parameters could be reduced to two, namely the average peak noise level of the aircraft and the number heard per day.

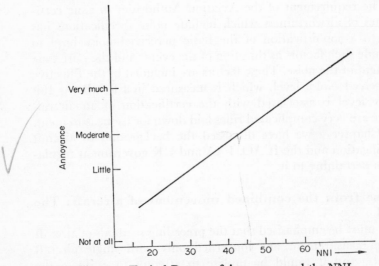

Fig. 5.2   Typical Degrees of Annoyance and the NNI

Furthermore, it was found that the average annoyance expressed could be considered to be a function of one composite variable, and that this could be taken as being the sum of the average peak noise level in PNdB and 15 times the Log N., where N is the number of aircraft heard per day. The composite variable was christened the Noise and Number Index, and eventually defined as:

NNI = The average Peak Noise Level $+15 \log N -80$

The -80 term was introduced to simplify the zero position of the scale. Using this number the range of reactions is shown against the value of NNI in Fig. 5.2 opposite.

The significant point of the development of this Index lies in the concept of adding up the individual events of noise to arrive at an average daily dose. The difference between this concept and that of the individual aircraft noise level is evident.

The limitation of the NNI

It is not always scientifically acceptable to relate data gained from one location to another situation. Neither is it always acceptable to extrapolate data beyond the values which were originally measured. The NNI is of such potential value to both Airport and Planning Authorities that on occasions it is used indiscriminately. For example, the relationship between NNI and subjective response has been used to assess the effect of short runways at small provincial airports in the UK. The differences in the type of aircraft involved, the levels of air traffic and the history of noise exposure of the populations involved may be very large, often large enough to give serious cause for doubt whether the reaction of the populations would be at all similar at a given level of NNI exposure.

There is also the argument that a population will become more used to noise over a period of time. This last point was investigated by a second survey of the effect of aircraft noise around London (Heathrow) Airport carried out for the UK Board of Trade in 1967. This work set out to repeat as far as was possible the work of the Wilson Committee and also to investigate in detail the effects of nighttime noise and the precise relationship between NNI and the subjective response of the population. The work was published in 1971 under the title *Second Survey of Aircraft Noise Annoyance Around London (Heathrow) Airport* (Ref. 28).

The report did in fact recommend minor changes in the method of calculation of the NNI level, but it was not considered to be of value to change the already well-established methods for the sake of a small increase in the accuracy of the correlation between NNI and annoyance. Unfortunately the second survey did nothing to clear up the doubts still felt concerning the effects of nighttime exposure to aircraft noise.

The thinking on the subject of the assessment of aircraft noise annoyance can be summed up by saying that the subjective reaction of an exposed population depends upon the total dose of noise which it receives in a given time. The definition of this total dose varies with different workers, but the variations are small. Work on this subject is of course not confined to the UK but is virtually world-wide. America, Germany, France and many other countries are involved in the study of what is essentially an international problem.

Noise assessment in the USA

The history of the evolution of the assessment of aircraft noise in America is interesting. As long ago as 1953 Rosenblith and Stevens introduced the concept of the Composite Noise Rating (CNR). Basically this Rating was defined by a single number which related to the intruding noise of the aircraft. To this basic number were added increments depending upon certain factors. These were four in number: firstly, repetitiveness, or the number of times a day the noise occurred, in fact corresponding to ten times the log of the number; secondly, background noise, an addition of 5 dB being applied for quiet suburban areas varying down to as much as a reduction of 15 dB for a highly industrialised area; thirdly, an allowance for nighttime disturbance; and fourthly, a final factor based upon the degree of community adaptation; this varying, rather at the intuition of the investigators, from five to ten units.

The relationship between this final composite number and the community response was investigated by examining a number of case histories of known community reaction to noise from aircraft. The basic concept was modified in 1955 to allow an extended range for background noise and to take a more detailed account of the duration of each individual noise event. An addition was made to allow for a seasonal correction. Further case histories were investigated and the use of the CNR gained acceptance as a method of predicting the likely change in public reaction following modification to an airport facility, or its method of operation. The CNR method received further investigation by the work of Borsky in 1961. Ultimately the method has undergone further refinements and has been used widely by the USAF for planning purposes at its airfields around the world. The Federal Aviation Agency did not accept the 1964 CNR Index, and sponsored in its turn two parallel projects to develop a method for making noise exposure forecasts at commercial airports in the United States. One study was conducted by Bolt, Beranek and Newman, and the other by an independent committee of the Society of Automotive Engineers. The two studies came up with virtually identical procedures which resulted in the Noise Exposure Forecast (NEF) method.

$$\text{NEF} = \text{Average noise level in EPNdB} + 10 \log N - K$$
$$\text{where } K = 88(\text{daytime } 0700\text{--}2200 \text{ hrs})$$
$$\text{or} \quad K = 76(\text{nighttime } 2200\text{--}0700 \text{ hrs})$$

No new community data was produced to support the NEF method and the appropriate community responses were derived by comparison with the CNR method. The NEF concept is now used by the FAA for most of its noise planning activities. ICAO has studied the problem of airport noise assessment at some length and recommend the use of indices which follow very closely the fundamental principles of the methods described in this chapter.

Between 1964 and 1967 other national groups also carried out studies and indices of annoyance assessment were evolved.

Other national methods of assessment

In Australia Murray and Piesse (1964) published an analysis of complaints from the area around Sydney International Airport and defined an annoyance index:

$$\text{A I} = \text{Average noise level in PNdB} + 10 \log N$$

The German workers have produced an index termed the Storindex $\overline{Q}$ which is defined as:

$$\overline{Q} = \text{Average noise level in}$$
$$\text{dBA} + 13\cdot3 \log N + 13\cdot3 \log \bar{t}/T$$

This index is interesting in so far as a term is included for the average duration of each noise event ($\bar{t}$). T is the reference time of the study. The French study of Paris Orly, Paris Le Bourget, Marseilles and Lyon (Coblenz and others 1967) came up with yet another index:

$$R = \text{Average Noise Level in PNdB} + 10 \log N - 30$$

In essence all these indices of 'annoyance' are similar in that they attempt to relate the expected subjective response to the quantity of noise which is received in a certain reference time. The exact definition of the quantity does vary somewhat from one research worker to another, but essentially it means multiplying the average noise of the aircraft by the number of aircraft heard, but what is probably of more importance in the use of any of these indices is to ensure that the relationship of the index to the subjective scale is relevant. It is not surprising that these various national indices show some variations in their definition as it will readily be realised that different ethnic groups will react to noise in slightly different ways.

Predictive techniques for planning

Planning authorities have recognised the need for the use of some predictive index in the work associated with the environs of airports. It is of immense value to them to be able to judge the expected community response from people in a certain area to the aircraft noise, and to weigh up the consequences of increasing the potential annoyance in that area following an increase of population. Conversely, areas unsuitable for residential development can be identified. Airport authorities can also be aided by these indices, since by the construction of contours to illustrate any proposed change in airport operations, the consequences of such actions can be demonstrated.

We have already discussed in Chapter 2 how the demands made by planning and airport authorities on these indices are becoming more and more severe. It is increasingly necessary to prove the validity of the index in a specific context. The County Planning Authorities of East Sussex, West Sussex and Surrey County Councils, for example, were aware of this problem as it affected their development control policy for the area around Gatwick Airport. This policy was eventually based upon the NNI concept and it utilised the subjective scale of annoyance derived in 1963 by the Wilson Committee.

The Wilson work was based entirely on Heathrow conditions and it is reasonable to suppose that the response of the population recorded by the social survey was coloured by factors of this environment. These factors may have had nothing to do with aircraft noise but could still effect the reaction to the noise. Such things as the socio-economic make-up of the population and the previous history of its exposure to noise could well have given rise to responses which would not have been forthcoming in a different location even if that location were exposed to a similar level of aircraft noise. Another factor which could have been of great significance was the amount of noise present which was not attributable to the airport. The area around Heathrow is intersected by two main road systems and the area itself is fairly industrialised. This intrusive background noise level would obviously affect the degree of additional intrusion imposed by noise from aircraft.

The importance of these factors was sufficient to cause the planning authorities around Gatwick to require some information on the actual effect of aircraft noise on the population, and not to rely entirely upon the transference of information from the Heathrow case. They wanted, in short, to ascertain if there were a difference in the response of the Gatwick population, and if so they wished then to be in a position to take this into account in their control policy for residential development.

The Gatwick social survey

In order to be in a position to be certain that their policy was soundly based the three county councils commissioned an investigation of the Gatwick area. The brief for the study was to repeat the work of the Wilson Committee and to define a scale of subjective reaction to aircraft noise which was relevant to the Gatwick situation. The social survey was carried out during the summer of 1971, and used the same basic questionnaire as that of the Heathrow surveys. At the same time the operations of the airport were subjected to close scrutiny and very detailed information describing the type, numbers and flight paths of the aircraft was gathered. These two sets of data, the subjective responses and the physical noise data, were used to construct a scale of annoyance against the NNI. No attempt was made to modify the concept of NNI as we have seen that only marginal improvements were possible in the correlations between subjective response and exposure to noise.

An advantage of this approach is also that the subjective response could be directly compared with that of the Heathrow population. Fig. 5.3 shows the condensed results of the study,

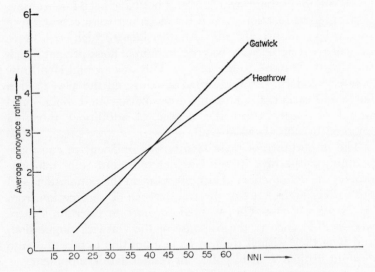

Fig. 5.3   Comparison of Gatwick and Heathrow Response

with the Heathrow values included for comparison. It can be seen that at about 40 NNI the response of the two populations is similar. At higher values of NNI the Gatwick response is more severe than the Heathrow response while at values lower than 40 NNI the Gatwick response is less severe than the Heathrow response. This information has been of great value to the three councils in that it enabled them to revise their development control policy to fit the conditions under consideration. Table 5.1 below shows the two policies for comparison.

The Government has seen the value of the use of a general policy for the planning of areas around the sources of noise and their circular 10/73 'Planning and Noise' published by the Department of the Environment has become the basic national standard. The section dealing with airport noise utilises the old Surrey County Council policy, based on Heathrow data, but does allow individual authorities to set different standards should local conditions so dictate (Ref. 14).

TABLE 5.1. *Revision of the Gatwick Residential Development Control Policy*

Daytime NNI Band		Planning Action
Former Policy	New Policy	
Above 60	Above 55	Refuse all major residential development
40–60	40–55	Refuse all major residential development. Permit infilling only with insulation against noise.

The problems discussed up to this point are common to all aircraft and airports, but there is one problem which is reserved to particular types of aircraft and furthermore is a problem which is not associated only with airports. It is also linked with the problem of operating supersonic transport. Mention has already been made of the difficulty in the development of a quiet engine capable of supersonic performance, but of even more importance is the question of the sonic boom.

Sonic Boom

The phenomenon which has been termed the sonic boom is created by a body moving through the atmosphere at a speed greater than the speed of sound. The mechanism of the boom is interesting and is worthy of some explanation.

When a subsonic aircraft moves through the atmosphere the air in front of the leading edge of the aircraft has sufficient warning of the approach of the aircraft to part smoothly and to flow over the upper and lower surfaces of the wing. This warning is not received in the case of an aircraft travelling at supersonic speeds, since the aircraft is travelling faster than the rate of warning of the approach. As a result, the air in front of the wings is not able to separate and flow smoothly over the aerofoil, but is piled up until the pressure is such that it parts and flows over the wings. A similar action takes place at the trailing edge and there is a similar pressure change as the air-flows join up again. Thus if the pressure were measured from just in front of the leading edge, across the wing chord to a point just past the trailing edge two large jumps would be observed, an increase of pressure at the leading edge, and a decrease of pressure at the trailing edge. It is these two pressure changes which give rise to two waves of disturbance or 'shock waves'. These waves travel in a cone from the aircraft much in the manner shown in Fig. 5.4

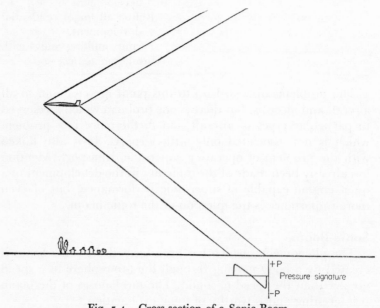

Fig. 5.4   Cross-section of a Sonic Boom

Where this cone impinges upon the ground the effect on an observer is for him to experience the double bang characteristic of the sonic boom. The levels of energy experienced in a sonic boom are too high to measure by the familiar methods of dBA or PNdB, but are expressed as an over-pressure, and the in-cluded angle of the cone depends on a number of factors, such as the height and weight of the aircraft and the acceleration.

Smicboom A

Whether or not the shock waves strike the ground depends on atmospheric factors, such as changes in air density, temperature and pressure. These can alter the shape of the waves in such a manner as to bend them away from the ground altogether. The width of the boom carpet is also dependent upon aeronautical and atmospheric factors (Ref. 5).

It should be remembered that the sonic boom effects only occur when an aircraft is travelling at supersonic speeds and that when operating subsonically the mechanisms of noise emissions obey the same rules as any other subsonic aircraft. This defines the sonic boom problem as one associated not with airports and the approach and departure patterns, but rather with aircraft flight corridors. Owing to the design characteristics of current supersonic aircraft such as the Anglo-French Concorde and the Soviet Tupolev 144, a distance of 70–100 miles is overflown after take-off at subsonic speeds during the acceleration phase up to supersonic cruising flight. A corresponding distance is flown subsonically before landing. Thus the problem is not of significance in the airport environment.

The width of the boom carpet can be of several miles and the number of people potentially affected by the supersonic boom over land could be immense. It does emphasise, moreover, that the sonic boom transfers the problem of aircraft noise from the airports themselves to the general population over a wide area. Indeed, it presents a problem to some populations which previously were only aware of the presence of aircraft by observing the high altitude condensation trails over their country.

The subjective reaction to sonic boom is quite different from subsonic aircraft noise. The character of the phenomenon is that no warning of the approach of the boom is given, consequently 'startle effects' may be very large. The effects of the boom on a structure may be observable, for example a large plate-glass window could visibly bulge under the effects of a sonic boom. This could be very alarming even though the window would be unlikely to break. The character of the boom from a supersonic transport is different from that typical of a fighter aircraft to such a degree that actual damage to any structure in a reasonable state of repair is much less likely to occur. Some work carried out at Truro Cathedral during the test runs of Concorde down the western coast of Britain showed, however, that more damage to the structure was caused by the playing of the organ than by the sonic boom from the aircraft.

However, world experience of the effects of sonic booms caused by transport aircraft is at present necessarily very limited, and the principal nations of North America and Europe are taking stringent precautions, having in many cases already banned supersonic flights over land by military and transport aircraft.

This may limit the commercial viability of this type of aircraft which will tend to concentrate on the long distance over-water routes of the world during the early stages of development. This, however, will leave open many of the important trunk routes, including the North Atlantic and Pacific ocean routes for economic exploitation in the 1980s.

6   Airports and Society in the 1980s

'An uninterrupted navigable ocean, that comes to the threshold of every man's door, ought not to be neglected as a source of human gratification and advantage.'—Sir George Cayley (1773–1857)

It is intended to bring together in this final chapter the main threads of the previous discussion so as to reach some conclusions on the prospects for environmentally orientated airport development, and the alleviation of impact from air traffic on the community in the 1980s.

Criticism and confidence

The airport has become a centre of conflict at several levels, but principally in the national press where it has been found to be a ready subject for controversy, and where the views expressed are almost wholly against major national development programmes; and secondarily in the local press where regional and local airport plans are so often quickly and uncritically condemned.

Such criticism holds within it a large amount of contradiction, and no man who is dependent on an industry with a significant international element could surely decry a policy which encourages air transport development, or any other system of transportation on a national scale. That such a system could be directed away from one's own residential area may be desirable but morality must surely play a part here, and so long as the Government and the authorities concerned can be shown to have used justice in their selection of options for development, and have employed the best independent data and advice in the assessment of the problem, the decisions of Ministers should be respected. It is a matter of confidence. Such confidence should be reinforced by the publicity given to government action to promote research into quiet engine technology, and to develop new transport aircraft. It should surely be amplified even more by a realisation, of the international action by ICAO, by IATA, by the World's aircraft industry and by the international airlines which operate through our major airports. Few industries can act as fast as the aircraft and engine manufacturers and the airline operating industries. The crises of the late twentieth century have demonstrated this. The public is generally inadequately informed on the noise and pollution reduction programmes in hand, and the results being achieved.

The fuel crisis

The fuel crisis, whose full extent it is most difficult to assess today, is unlikely to enforce any permanent curtailment of the programmes which we have described in these pages. That the present inflationary trends in wages, in air traffic services of all kinds and in fuel will have a serious short-term influence upon growth must now be accepted. To this the fuel supply restriction will add perhaps for some years a constraint upon potential expansion until the Middle Eastern States can find a long lasting political settlement. However, short of a major worldwide economic crisis it is quite reasonable to presume that the Government and industry will be equal to the problems set by inflation and fuel supply, and that noise and pollution in the environment will be safeguarded by wise control of progress rather than by the stagnation of trade and transport.

Air safety

But other adverse aspects of air transport must be contained. Not only noise and atmosphere pollution can inflict damage on men. The fear of disaster is present in the minds of many who live in an airport environment. This anxiety can only be allayed by the continuous improvement in the safety record of the air transport operating industry. With no improvement in the accident rate per million flying hours of civil aviation we shall be exposing more passengers to the risk of accident. Thus we shall need to reduce the accident rate much further with the larger aircraft which are coming into service in spite of the apparent improvement in the level of fatalities per passenger-mile.

The role of local authorities

The control of these problems may be largely in the hands of central governments, but the implementation of much of the environmental policy is handled by local authorities.

These are the landed gentry of the late twentieth century and in the last two decades have undertaken sterling work in the control of land-use and in manifold aspects of environmental protection. National airport policy is generally beyond the survey of local authorities, but their responsibility for their own communities will reach out into the national air traffic issues when noise from overflying aircraft reaches the magnitude of a serious environmental problem – or when a great airport is developed or is newly sited near to or within their boundaries.

Like all public authorities, they have recently become more expert in the art of public relations, and public participation is now a cornerstone of much of the planning policy of local and regional government in most Western countries. Probably far more could be done in this direction in general. In particular, the author believes that a far greater amount of information on airport planning and development policies should be published. So much criticism stems from an ignorance of the facts, and half-understood schemes have been distorted by the well-intentioned with a resulting increase in concern and opposition to sound projects.

The economic factors

The significance of air transport as a form of commercial enterprise has now been firmly established, and in all highly developed countries it has become accepted as second only to the motor car as the most used form of inter-city travel. Its wide scope is emphasised by its provision of services in the business, private and tourist sectors and in its dramatic growth since the 1950s. Familiarity in fact has led to some contempt, and the air journey is now rarely undertaken for pleasure. In the less fully developed countries air transport has also proved to be an important source of foreign exchange and without exception an essential link with the outer world of politics and trade. It is hard indeed to find a country able to neglect without economic loss its international air services and the infra-structure, including airports which support them. Domestic air services have often been found more difficult to justify unless the significant journeys exceed 200–300 miles.

To the aeronautical industrial countries such as the USA, Canada, the UK, Japan and France the significance of air transport is even wider. World markets for civil aircraft, engine and avionics have become important factors in the balance of payments of special significance to the USA, as to the UK, each with its challenging trade balance problems.

The impact of the jet aeroplane has created the principal forms of objection to the growth of civil aviation. The extensive introduction of jet services in the 1960s was at the same time the instrument for world air carrier expansion and for the concentration of noise in the environs of the major world airports, which grew in size and extent to handle the big jets. It was the increase in demand for runway extension, the growth in air traffic and the noise increase of the high-thrust engines required which created the world airport noise problem. Here arose the first conflicts between air transport and the community.

Air traffic and noise control

Pressure by local communities reacting to aircraft noise has now indeed become an important feature in airport planning. Recent policy decisions by the UK Department of the environment have emphasised the need for airport authorities to make careful examination of noise problems, and to avoid any immediate development which may at a later date lead to constraints being imposed. The Civil Aviation Act extends government powers over the control of aircraft noise to all airfields licensed for public use. Responsibility for meeting the requirements imposed by government will rest with airport authorities. In the USA the Airport and Airways Development Act, 1970, established a requirement for noise nuisance to be minimised by site selection and abatement procedures adequate to meet community needs.

Many airport authorities therefore find themselves much involved in aircraft noise problems and their alleviation on and off the airport.

The introduction of new aircraft types will effect a major reduction in noise levels at the source, and this should be taken into account when establishing any noise limiting policies for future years. In particular, conditions and procedures which are excessively restrictive, and incapable of taking these developments into account, should be avoided.

The introduction of quieter aircraft

The extension of European operations with wide-bodied jets and notably the McDonnell-Douglas DC 10, the Lockheed Tristar (and later with the A-300B European Airbus) will significantly reduce flyover noise levels. Already important improvements have been recorded in the USA and the UK.

Noise certification, applicable to new jet transports with few exceptions, operating in the UK and in Western Europe will ensure that future airline fleets achieve lower noise levels. Non-transport aircraft, in both the business executive and private spheres are also becoming quieter, and it seems highly likely that noise certification for these types will also be called for by the end of this decade.

Further technological innovation

We should note, however, the important contribution made by the steadily increasing size of transport aircraft.

Aircraft movement rates at nearly all major transport airports have increased far less rapidly than that of the passenger and cargo themselves. With the lower noise levels of new generation aircraft this factor is already becoming significant in its influence on noise restraint through the 1970s.

In the current state of the art, there are rather limited means available for keeping loaded aircraft away from people in the immediate vicinity of airports. The most powerful means of achieving this remains largely in the long-term future when vertical take-off and landing (VTOL) and (perhaps rather sooner) short take-off and landing (STOL), should become commercially acceptable for large-scale air transport operations. These solutions will be found first of all in short- and medium-haul sectors rather than in long-haul inter-continental type operations. Interim alleviation may be found in the new short-medium haul aircraft almost certainly coming into service in the last years of this decade. These aircraft with reduced take-off and landing run requirements will introduce a new interim

formula for air transport with exceptional features favourable
to the environment.

Siting and noise abatement

The geographical situation of the airport and the surrounding
community is often such that the selection of a particular route
for aircraft, or the preferred use of a runway direction, will
considerably reduce the impact of aircraft noise. Often this can
be achieved without imposing any cost penalty upon the operation of the aircraft or without employing the reduction of power
that is called for by noise-abatement climb procedures. The
types of aircraft likely to use an airport has, of course, considerable effect on the potential and this is especially significant today
with the imminent introduction of new types of quieter aircraft
into airline fleets.

Alternative noise minimisation procedures should be an
integral part of any noise reduction programme. These would
include not only flight procedures in the immediate vicinity of
the airport, but air routings over a far wider area when significant to the community.

The location and orientation of aircraft facilities can also
have valuable benefits in improving relations with the local
community. A careful examination of the siting of apron areas,
taxiways, engine run-up areas, cargo bases and maintenance
facilities will often indicate an optimum solution which can
reduce aircraft noise in relation to residential concentrations,
whilst retaining operational viability.

Public participation

More extensive use of the press and other media of the
national community, as well as professionally organised groups
of objectors, have greatly changed the situation in official
public inquiries and in the unofficial debates when major
projects of all kinds are under scrutiny.

Within the framework of UK planning law the specialist
professions have been rallied to the cause. Participation by the
community in the decision-making process must now be
accepted by local authorities and government departments up
to the point at which a final decision between various conflicting
views needs to be made. This democratic process may be

thought to be far more autocratic than it is made to appear, but in our view this is essential to strong policy making.

In any case, aviation introduces so many factors demanding inputs from so many disciplines that these are rarely able by themselves to provide definitive and agreed answers.

Cost-benefit analysis on its own can only carry the numerical assessment up to a certain stage where sound judgement based on a careful compilation of the available facts must always take over. Political decisions may then prove to be essential, and generally these are the last words, though we may often have to wait a long time for them.

The opposition to airport development generally evolves in two ways – firstly, the altuistic effort to maintain the environment in all respects, and against all-comers, with the aim of minimum pollution on every count; and secondly, the personal aims of like-minded people who wish to maintain their own domicile and its surroundings to an existing standard, and especially to sustain its value because their property represents in economic terms their most important single possession. These two aspects of opposition to airport development are usually confused, and sometimes with intent. It is very essential that they should remain separate and be recognised as such.

Progress and economic compromise

In most countries the areas of land available in the environs of the major cities is steadily decreasing as the pressure of demand for residential estates and industry rises fast, and will probably go on rising during the foreseeable future. Thus the need to resolve the problems we have been discussing is likely to become more acute.

Who should be more capable of solving the outstanding problems than the aviation industry backed by governments and the international agencies. We have tried to describe the principal lines of such activity in this book. Concerted efforts in the USA and in Europe must surely find significant solutions within two decades, and many have already been found.

But progressive men and lovers of our lands alike will not remain content until the noise from all forms of transport and industry has been beaten down into the ground. In that day

we shall hear the bird song and the cicadas, and need to make
no such tortuous journey as we have made in these pages
around government, the community, the airport and the
fixed-wing transport aeroplane. Other problems and conflicts
will assuredly appear to confuse us then.

'Now earthward in their main intent
. . . . soon,
Minded beyond the moon
Man will enlarge his winged experiment.'
—Edmund Blunden.

References

1. *The Third London Airport*, Report of the Roskill Commission, 1971.
2. *Civil Aviation Research and Development Policy Study*, DOT and NASA, Washington, 1971.
3. Reports of the Air Transport and Travel Industry Training Board, 1972–4.
4. Reports of the Society of British Aerospace Companies, and of the Aerospace Industries Association of America, 1965–73.
5. *Air Transport Economics in the Supersonic Era*, 2nd ed., Macmillan, 1973.
6. 'Demands for Air Transport, 1980–90', K. Hammarskjöld, Airports Conference, London, 1973.
7. *The Importance of Civil Transport to the United Kingdom Economy*, IATA, 1970.
8. *Review of the Economic Situation in Air Transport*, ICAO, 1971.
9. *The Benefits of Civil Aviation to the Jordanian Economy*, The Royal Jordanian Airline, Alia, Amman, Dec. 1972.
10. 'Air Service Development in Third World Countries', International Business Communications Conference. A. H. Stratford, 1972.
11. *Airport Planning. An Approach on a National Basis*, CAA, Dec. 1972.
12. *Noise. The Social Impact*, W. Allen.
13. 'Proposed Zones for the Control of Development: Gatwick', Surrey County Planning Dept, 1968.
14. 'Planning and Noise', Dept. of Environment Circular, 10/73.
15. *Aircraft Noise. Flight Routing near Airports*, HMSO, 1971.
16. *Assessment of Noise at Various Stages of Airport Development*, Fleming.
17. 'Contrasts in National Airport Planning', R.P.A. Roos, Airports Conference, London, 1973.
18. 'Future Aircraft Noise in the Vicinity of Airports', GEERT Zimmerman, Airport Forum, 3/1971.
19. 'Contrasts in National Airport Planning', J. H. Shaffer, Airports Conference, London, 1973.

20. ICAO, Annexe 14, 'Aerodromes'.
21. ICAO, Annexe 16, 'Aircraft Noise'.
22. *The Licensing of Aerodromes*, CAP 168. HMSO, 1969.
23. *The Protection of the Environment*. Cmnd. 4373, HMSO, 1970.
24. *The State of the Air Transport Industry*, IATA, Montreal, 1972.
25. 'Quiet Aircraft', J. T. Stamper, Local Authorities' Aircraft Noise Conference, 1972.
26. 'The Quiet Side of NASA', *Flight International*, 6 July 1972.
27. *Noise*. Report of the Wilson Committee, Cmnd. 2056, HMSO, 1963.
28. *Second Survey of Noise Annoyance around London (Heathrow) Airport*, HMSO, 1971.
29. 'Nature and Composition of Aircraft Engine Exhaust Emissions', Report 1134/1, Northern Research and Engineering Corporation, 1968.
30. 'Air Pollution at Heathrow Airport', J. Parker, Department of Trade and Industry. SAE/DOT Conference, 1971.
31. *Action against Aircraft Noise*, Annual, HMSO, 1973.
32. *The Future of Aeronautics*, ed. John Allen, Royal Aeronautical Society.
33. 'Noise and Society', E. J. Richards, *Journal Royal Society of Arts*, 1971.
34. 'Aviation and the Environment', P. G. Masefield, *The Aeronautical Journal*, 1971.
35. *Supersonic Transports and the Environment*, BAC, Weybridge, A/3/72.
36. 'Quiet Aero Engines Cause & Effect', M. J. T. Smith, British Acoustical Society, Spring Meeting, 1973.
37. *Aircraft Noise in the Vicinity of Aerodromes*, ICAO doc 8857, 1970.
38. *Assessment of Aircraft Noise and Community Response*, Stratford and Waters, British Acoustical Society, 1973.

Index

House values, 59–61
Housing near Heathrow, 47, 48
Hush kits, 64, 99–101

IATA, 64, 69, 97, 99
ICAO, 7, 16, 39, 62, 64, 78, 79, 80, 95
Independent aviation authorities, 41–3
Instrument runways, 80
Insurance, revenues, 26
Intensity of noise, *see* Noise levels

Japan R&D, 124
Jet aircraft fleets, 9, 112
Jet efflux noise, 110, 129
Jet engine development, 10, 99–103, 130
Jordan, 31
Journey times, 58, 59

Landing noise, 104, 105
Licensing of aerodromes, 80
Lighthill, Sir Charles, 109
Location of facilities, runways, 84, 90
Location of monitoring, 49
Lockheed Tristar, 105, 111–12
London Airport design competition, 1
London Airport, pollution, 93–4
London Airport, site, 1
Los Angeles Airport, 28, 74

McDonnell-Douglas aircraft, 112
McDonnell-Douglas DC 8 – 9, 72, 120
McDonnell-Douglas DC 10 – 105
Manchester Airport, 89
Manchester Corporation General Powers Act, 89
Maplin Airport, 6, 75, 87
Master plans, 57, 81
Massachusetts Port Authority, 90
Minimum noise routes, 52, 53
Monitoring of noise, 48–50, 150
Motivation for air journeys, 18, 19
Motives for objections to noise, 13

NASA research, 122
NASP, 73, 74
NEF, 137
Net overseas income, 24
New aircraft types, 117, 149
Night curfews, 87–9
NNI, 125, 133–6
Noise abatement, 44, 103–4, 150
Noise Advisory Council, UK, 53
Noise and Number Index, *see* NNI
Noise certification, 62
Noise control, 44, 48, 49, 87–9, 150
Noise Exposure Forecast, *see* NEF
Noise impact, 10, 116, 148
Noise insulation, 44, 45, 48, 89, 90
Noise levels, 10, 48, 50, 111–13, 120, 132, 133
Noise measurement, 131
Noise monitoring, 48–50, 150
Noise nuisance contours, 46, 125, 133, 137
Noise nuisance measurement
 Australia, 137
 France, 138
 UK, 133–6
 US, 136–7
 West Germany, 137
Noise principles, 127
Noise shielding, 121
Noise standards, 62, 106
Non-Instrument Runways, 80
North Atlantic, air penetration, 24

Off-shore sites, 75
Older type jet movements, 106
Other noise, 50
Overseas income, 24
Overseas visitors, 23

Palmdale, 28, 74
Paris exhibitions, 34
Parliament, part played by, 6
Penetration of air transport, 24
Perceived noise level, 132
Planning systems, English, 6
PNdB defined, 132
Political factors, 35–7
Pollution, atmospheric, 92–5